Lecture Notes in Operatio

Lecture Notes in Operations Research is an interdisciplinary book series which provides a platform for the cutting-edge research and developments in both operations research and operations management field. The purview of this series is global, encompassing all nations and areas of the world.

It comprises for instance, mathematical optimization, mathematical modeling, statistical analysis, queueing theory and other stochastic-process models, Markov decision processes, econometric methods, data envelopment analysis, decision analysis, supply chain management, transportation logistics, process design, operations strategy, facilities planning, production planning and inventory control.

LNOR publishes edited conference proceedings, contributed volumes that present firsthand information on the latest research results and pioneering innovations as well as new perspectives on classical fields. The target audience of LNOR consists of students, researchers as well as industry professionals.

This series is indexed by Research Papers in Economics (RePEc), zbMATH and has been accepted by Scopus.

Filipe Alvelos · Isabel Martins ·
Ana Maria A. C. Rocha
Editors

Advances in Optimization and Wildfire

Proceedings of the First Optimization and Wildfire Conference, October 1-4 2024, Luso, Portugal

Springer

Editors
Filipe Alvelos
Department of Production and Systems/
ALGORITMI Research Centre/LASI
School of Engineering, University of Minho
Braga, Portugal

Isabel Martins
School of Agriculture/Center for
Mathematical Studies
University of Lisbon
Lisbon, Portugal

Ana Maria A. C. Rocha
Department of Production and Systems/
ALGORITMI Research Centre/LASI
School of Engineering, University of Minho
Braga, Portugal

ISSN 2731-040X ISSN 2731-0418 (electronic)
Lecture Notes in Operations Research
ISBN 978-3-032-03107-5 ISBN 978-3-032-03108-2 (eBook)
https://doi.org/10.1007/978-3-032-03108-2

This Springer imprint is published by the registered company Springer Nature Switzerland AG
The registered company address is: Gewerbestrasse 11, 6330 Cham, Switzerland

Foreword

In 1961 Shephard and Jewell identified fire management as a topic to which they thought operational researchers could and should contribute their expertise. A small number of operational researchers subsequently followed their lead and published their findings in operational research (e.g., Operations Research and European Journal of Operational Research), forestry (e.g., Canadian Journal of Forest Research and Forest Science) and fire management (e.g., International Journal of Wildland Fire) journals.

This book contains 15 chapters, each of which is based on a paper that was presented at the Optimization and Wildfire Conference that was chaired by Prof. Filipe Alvelos (University of Minho, Portugal) and co-chaired by Prof. Isabel Martins (University of Lisbon, Portugal) and Prof. Ana Maria A. C. Rocha (University of Minho, Portugal). The conference, that I was fortunate to have attended, took place in Luso Portugal, October 1–4, 2024.

There were a total of 58 participants from 14 different countries and 46 presentations (authored by 135 contributors) that addressed a broad range of topics including fuel management, detection, suppression resource deployment and dispatch, the protection of values-at-risk, fire management budgeting, fire growth simulation, risk analysis, remote sensing and the development of decision support systems. A total of 15 papers based on those presentations are published in this book.

There were three keynote speakers at the conference of which I was privileged to have been one. The other two were Yu Wei from Colorado State University and José Borges from the University of Lisbon.

Chapter 11 is based upon my presentation during which I discussed wildfire management from an operational research perspective, described some of the research on which my colleagues have focussed our attention in recent years, and identified some open problems to which I believe operational researchers can make important contributions.

Yu Wei focussed on two very important topics, spatially explicit modeling and managing uncertainty. He focussed on the use of wildfire decision support systems to address spatially explicit fire risk modeling during which he discussed the use of raster and vector approaches, how to encode uncertainty, combinatorial optimization challenges, fuel treatment planning, initial attack planning, and the impact of fire on watersheds.

José Borges focussed on the increasingly important need for integrated fire and forest management and the importance of addressing ecosystem services under global change. He described some of the major planning and modelling challenges he and his colleagues have addressed and how they did so across a broad range of spatial and temporal scales and how they dealt with the uncertainty that complicates such planning and decision-making when they did so. He also discussed the importance of working with stakeholders and how to address their needs and the design of strategic networks of managed areas to improve suppression efforts against Extreme Wildfire Events (EWEs).

Mauricio Acuna et al. (Chap. 1, Stochastic Optimization for Scheduling Thinning Operations in Mediterranean Pine Forest Stands Under Fire Damage Risk) develop a two-stage stochastic programming model to optimize a thinning program over a 10-year planning horizon that maximizes Net Present Value by incorporating revenues from biomass, timber sales, and carbon credits, while minimizing fire risk, represented by the fire suppression difficulty index.

Alan Ager and Hugh Safford (Chap. 2, Advances in decision support platforms for prioritizing investments in forest and rangeland restoration, risk reduction and biodiversity conservation) reviewed selected landscape planning and prioritization platforms (integrated solutions that leverage one or more technology building blocks including big data, spatial optimization algorithms, geospatial technology, and cloud computing) and their application in diverse ecosystems and operational environments. They concluded that the uptake of these newer platforms and underlying science by public and private entities is not keeping pace with the rate of technology development.

Filipe Alvelos (Chap. 3, pyO3F—A python framework for wildfire-related optimization, Part I: Design and fundamentals) describes the py03F framework he developed to support the development of optimization approaches including the use of meta-heuristics or mixed integer programming models, for wildfire-related optimization problems, that capture the interaction of suppression resources and growing fires on interacting networks. He followed that with (Chap. 4, pyO3F—A python framework for wildfire-related optimization, Part II: Usage) in which he describes how pyO3F can be used to address wildfire optimization problems.

Filipe Alvelos et al. (Chap. 5, A MIP model for wildfire extended attack) extended existing MIP models for fire spread and resources positioning and developed a mixed integer programming model that integrates fire spread and resource operations (movement, direct, and indirect attacks), both spatially and temporally address the problem of routing the firefighting resources available for extended attack on a wildfire.

Filipe Alvelos and Isabel Martins (Chap. 6, Integrating fire suppression in forest management) extended their earlier fire suppression research by integrating forest management and the placement of fire suppression bases in a single optimization model and they tested their approach by applying it to Portuguese forest.

Helena Alvelos et al. (Chap. 7, Modelling Wind Behaviour for the Development of Scenarios in the Context of Wildfire Spread) used an array of statistical methods to model historical wind data recorded at a Portuguese weather station such that it could be used to inform the pre-positioning and subsequent deployment of suppression resources on fires, the growth of which would be influenced by wind.

Miguel Constantino et al. (Chap. 8, Forest Road Network for Firefighter Access) address the need for cost-effective road networks that firefighters can use to travel to fires. They developed a Steiner Tree Problem model and present results from a case study conducted in a Portuguese forest.

Enrico Iurlano et al. (Chap. 9, The Graph Burning Problem under Constrained Diffusion) propose a Constrained Diffusion Graph Burning Problem that could be used to inform the management of fires, the spread of which is constrained by both natural barriers and suppression efforts.

Francisco Marques et al. (Chap. 10, A robust approach for the prepositioning of resources for wildfire suppression) developed a robust optimization model to deal with the uncertainty that complicates daily pre-positioning and subsequent dispatch of suppression resources to fires that may be reported later that day.

Alan Murray (Chap. 12, Optimization and Wildfire: Supporting Strategic Mitigation of Extreme Events) highlights the importance of spatial optimization and geographic analytics to support vegetation treatment planning.

Elsa Silva et al. (Chap. 13, Covering with Network Design for Wildfire Promptness) studied the problem of where to locate fire detection facilities and fire management bases and adopted a set covering approach and developed a mixed integer programming model to inform where to place such facilities so as to minimize the time required to respond to fires that occur on flammable landscapes.

Jessica Singer et al. (Chap. 14, An Integer Programming Model For Sensor Placement In LoRaWAN Networks) describe an integer programming model that was developed to inform the creation of a ground-based fire sensor network in a park in Buenos Aires, Argentina.

Sascha Zell and Armin Fügenschuh (Chap. 15, Optimizing Autonomous Unmanned Aerial System Deployment Locations for Enhanced Wildfire Detection and Monitoring) describe a location-allocation optimization model that can be used to inform the management of UAVs so as to minimize the time required for them to reach a new fire and guarantee a certain monitoring time at the scene.

Those chapters can be grouped into four broad categories that I often use to classify OR applications in wildfire management although some span more than one of those categories and the authors of some chapters may well take issue with my choice of categories and how I classified their contributions.

Strategic planning:
Acuna et al. (Chap. 1) and Murray (Chap. 12)

Presuppression planning including detection and deployment:
F. Alvelos and Martins (Chap. 6), H. Alvelos et al. (Chap. 7), Constantino et al. (Chap. 8), Iurlano et al. (Chap. 9), Marques et al. (Chap. 10), Martell (Chap. 11), Silva et al. (Chap. 13), Singer et al. (Chap. 14), and Zell and Fügenschuh (Chap. 15)

Suppression operations:
F. Alvelos et al. (Chap. 5) and Iurlano et al. (Chap. 9)

Development and adoption of decision support systems:
Ager and Safford (Chap. 2) and F. Alvelos (Chaps. 3 and 4)

What I heard and learned at the Luso conference bodes well for the future use of operational research to inform wildfire management as climate change, changes in land use practices and other factors continue to "ramp up" the threat that wildfire poses to people, their communities, their forests and wildlands and their other values at risk. What transpired in Luso indicates that;

- [1] vision has been embraced by many wildfire management agencies and operational researchers in many countries.
- Not surprisingly, their OR "toolboxes" have grown to include many powerful new tools that are being applied to both old and new wildfire management problems.
- Wildfire management problems and the need for operational research and related decision sciences are not going to go away anytime soon.

David Martell
Institute of Forestry and Conservation,
John H. Daniels Faculty
of Architecture, Landscape and Design
University of Toronto
Toronto, Canada

Reference

1. Shephard, R.W., and W.S. Jewell.1961b. Operations research in forest fire problems. In A study of fire problems. Nat. Acad. Sci. Natl. Res. Counc. Comm. Fire. pp. 145–165.

Acknowledgements

This book comprises 15 chapters of selected works presented during the Optimization and Wildfire (O&W) Conference that took place from the 1st to the 4th of October 2024 in Luso, Portugal.
First, we would like to acknowledge the authors of these chapters for sharing their knowledge and current research in such a relevant area. Their work is a valuable scientific contribution in its own and also in the establishment of optimization as a significant discipline in wildfire management.
A special mention is deserved to David Martell who, besides being an author, was also an invited speaker at the Conference, a reviewer and the author of the foreword of this book.
We are grateful to the reviewers for the quality of the comments and insights that led to the improvement of the chapters. Their excellent work was deeply appreciated.
Most reviewers were members of the Program Committee, Abílio Pereira Pacheco, University of Porto / ForestWise Lab, Portugal; Ana Maria A. C. Rocha, University of Minho, Portugal; André Mendes, University of São Paulo, Brazil; Andrea Trucchia, CIMA Foundation, Italy; Babak Abassi, RMIT University, Australia; Begoña Vitoriano, Universidad Complutense de Madrid, Spain; Fabricio Oliveira, Aalto University, Finland; Filipe Alvelos (chair), University of Minho, Portugal; Günther Raidl, TU Wien, Austria; Isabel Martins (co-chair), University of Lisbon, Portugal; Jaime Carrasco Barra, Universidad Tecnológica Metropolitana, Chile; Jose Ramon Gonzalez Olabarria, Forest Sciences Centre of Catalonia, Spain; Mahdi Bashiri, Coventry University, UK; Michael A. Forbes, University of Queensland, Australia; Miguel Constantino, University of Lisbon, Portugal; Mostafa Rezaei, ESCP Business School, France; Peng Wu, Fuzhou University, China; Sándor F. Tóth, University of Washington, USA; Susete Marques, University of Lisbon, Portugal; Woodam Chung, Oregon State University, USA, but we also relied on additional ones who deserve a special acknowledgement: Adelaide Cerveira, University of Trás-os-Montes and Alto Douro, Portugal; Agostinho Agra , University of Aveiro, Portugal; Alan Murray, University of California at Santa Barbara, USA; Ana Raquel Xambre, University of Aveiro, Portugal; David L. Martell, University of Toronto, Canada; Elsa Silva, University of Minho, Portugal; Helena Alvelos, University of Aveiro,

Portugal; João Pedro Pedroso, University of Porto, Portugal; Lino Costa, University of Minho, Portugal; Yu Wei, Colorado State University, USA.

This book would not have been possible without the success of the conference. We sincerely thank the 58 participants (46 presenters) and the indispensable efforts of the Organizing Committee of O&W:

Ana Maria A. C. Rocha (co-chair), University of Minho, Portugal; Ana Raquel Xambre, University of Aveiro, Portugal; António Bento Gonçalves, University of Minho, Portugal; Filipe Alvelos (chair), University of Minho, Portugal; Helena Alvelos, University of Aveiro, Portugal; Isabel Martins, University of Lisbon, Portugal; Lino Costa, University of Minho, Portugal; Marco Marto, University of Minho, Portugal; Senhorinha Teixeira, University of Minho, Portugal, who demonstrated remarkable dedication in planning and supporting all the activities of the conference.

In addition, we are grateful to the keynote speakers: David Martell (University of Toronto), Yu Wei (Colorado State University), and José C. Borges (University of Lisbon).

We deeply appreciated the presence in the opening session of Tiago Oliveira, President of the Portuguese Agency for the Integrated Management of Rural Fires (AGIF), and Pedro Arezes, Dean of the Engineering School of the University of Minho, Portugal.

We also acknowledge the support of **IUFRO** (International Union of Forest Research Organizations) and **APDIO** (Portuguese Association of Operational Research) in the dissemination of the Conference. We extend our acknowledgements to the **Portuguese Foundation for Science and Technology** that funded Project "O3F—An Optimization Framework to Reduce Forest Fire", involving the University of Minho, the University of Aveiro and the School of Agriculture of the University of Lisbon. Lastly, we acknowledge the support of the Research Centers: **Algoritmi**, School of Engineering, University of Minho, **CIDMA**—Center for Research and Development in Mathematics and Applications, University of Aveiro, and **CEMS.UL**—Center for Mathematical Studies, University of Lisbon.

Lastly, we would like to acknowledge the support of Springer. In particular, we thank Jialin Jin, Publishing Editor of Lecture Notes in Operations Research, for her professionalism and availability from the first contact. We extend our thanks to Sathya Subramaniam and Zeenathul Raeesa for their dedication in the production of this book.

Filipe Alvelos
Isabel Martins
Ana Maria A. C. Rocha

Contents

Stochastic Optimization for Scheduling Thinning Operations in Mediterranean Pine Forest Stands Under Fire Damage Risk

M Acuna, MA Varo-Martínez, V Lerma-Arce, G Palacios-Rodríguez, and R Navarro-Cerrillo

Abstract Forest planning in Mediterranean pine forests requires balancing the economic benefits of commercial thinning with the risks associated with wildfire suppression. This study applies two-stage stochastic programming (SP) to optimize a thinning program across a 45,000-hectare area over a 10-year horizon (Special Plan). The approach focuses on maximizing Net Present Value (NPV) by incorporating revenues from biomass, timber sales, and carbon credits, while minimizing fire risk, represented by the fire suppression difficulty index. The findings reveal that thinning operations and timber supply are significantly influenced by carbon prices and fire risk. The optimized thinning schedule prioritizes activities in the early and final years of the Special Plan, with reduced activity during the intermediate years. As fire risk increases, timber demand shapes thinning schedules, but this demand tapers off toward the end of the planning period. The study highlights the critical role of stochastic programming in addressing uncertainty related to fire risk. The low Expected Value of Perfect Information (EVPI) and high Value of the

---

The original version of the chapter has been revised. A correction to this chapter can be found at https://doi.org/10.1007/978-3-032-03108-2_16

M. Acuna (✉)
Natural Resources Institute of Finland (Luke), Helsinki, Finland
e-mail: mauricio.acuna@luke.fi

M. Varo-Martínez · G. Palacios-Rodríguez · R. Navarro-Cerrillo
Dpto. Forest Engineering, University of Cordoba, Córdoba, Spain
e-mail: mavaro@uco.es

G. Palacios-Rodríguez
e-mail: g52parog@uco.es

R. Navarro-Cerrillo
e-mail: rmnavarro@uco.es

V. Lerma-Arce
Institute of Information and Communication Technologies, Polytechnic University of Valencia, Valencia, Spain
e-mail: vlerma@upv.es

F. Alvelos et al. (eds.), *Advances in Optimization and Wildfire*, Lecture Notes in Operations Research, https://doi.org/10.1007/978-3-032-03108-2_1

Stochastic Solution (VSS) indicate that while knowing the exact fire risk would not dramatically improve the model's performance, incorporating uncertainty into the planning process greatly enhances silvicultural strategies. This underscores the need for adaptive forest management that leverages optimization tools and strategic planning to ensure sustainable timber supply, maximize carbon sequestration, and mitigate wildfire risks effectively.

Keywords Stochastic optimization · Thinning scheduling · Adaptive silviculture · Planning under uncertainty · Risk management

1 Introduction

Forest ecosystems contribute significantly to mitigating the effects of climate change, as the carbon (C) retained in phytomass represents the second-largest carbon reservoir (42%) of the total amount [8]. In Spain, this means that around 19% of total CO_2 emissions are sequestered by forests [9]. Silviculture optimizes the carbon sequestration capacity of forest systems by increasing stand growth, improving forest health, and enhancing productivity [19]. Significant advances have been made in biomass estimation and mapping for forest carbon calculations over large areas [12]. Similarly, advancements in remote sensing techniques and the integration of LiDAR data provide a new perspective on carbon inventories by directly offering three-dimensional information about areas under forest management [14].

Between the 1960s and 1970s, significant forest plantations were established in southern Spain, primarily for soil protection purposes. Specifically, in the Sierra de Los Filabres (Almería), these plantations involved various species, with *Pinus halepensis* Mill., *Pinus pinaster* Aiton., *Pinus nigra* Arnold., and *Pinus sylvestris* L. standing out due to their extensive coverage. These plantations resulted in highly homogeneous forests but also forests of considerable interest for forestry and restoration purposes. The forested areas within these plantations are subject to management plans aimed at preserving the health of the stands and, where possible, balancing public and conservation uses with generating direct benefits that offset silvicultural costs or provide direct income to the property, often municipally owned. However, one of the main risks to the stability of these stands is wildfires [10]. Wildfires are one of the most significant recurring disturbances in this type of forest system, with their frequency, intensity, and severity increasing over recent decades [4], a trend exacerbated by climate change [10]. In this context, silvicultural planning must integrate fire risk through both prevention and suppression efforts and the design and planning of silviculture, which remains a key issue in Mediterranean forest policies [21].

In this context, there are very few studies that analyze the optimization of silvicultural practices to maximize the Net Present Value (NPV) of interventions, enhance carbon sequestration capacity, and minimize fire risks. This complex problem can be effectively addressed using stochastic programming (SP). SP is an advanced optimization technique that incorporates probabilistic elements to model potential future outcomes. This approach allows decision-makers to plan for variability and adapt their strategies as uncertainties unfold [1]. SP typically operates within a two-stage decision-making framework. In the first stage, decisions are made based on available information before uncertainties are revealed. Once uncertain outcomes are observed, second-stage decisions, or recourse actions, are taken to adjust and optimize the initial plan. This structure enables SP to effectively address both immediate and future uncertainties [20]. A key feature of SP is its ability to model uncertainty through scenarios, each assigned a probability reflecting its likelihood of occurrence. This ensures robust and flexible solutions that account for a range of possible futures [22]. By integrating these probabilistic scenarios, SP not only minimizes risks but also enhances the adaptability and resilience of forest management strategies, providing a structured framework to tackle challenges under uncertain conditions.

The versatility of SP makes it an invaluable tool in forestry, capable of addressing a wide range of complex challenges. It supports sustainable forest conservation and utilization by providing models that balance resource use with biodiversity preservation under uncertain risks such as pests or wildfires [3]. Additionally, it aids in reducing wildfire risks by optimizing the allocation of resources for prevention and suppression efforts while incorporating uncertainties related to the location and severity of such events [18, 23]. SP facilitates optimal harvest planning by developing cutting schedules that maximize economic returns while minimizing climatic and ecological risks [6]. Furthermore, it enhances forest restoration efforts through models that account for climatic uncertainties and ecosystem responses, helping to determine which areas to restore and the most suitable species to plant [24]. This adaptability underscores its potential to drive innovation and resilience in forest management.

2 Objectives

The primary aim of this study was to apply stochastic programming (SP) techniques to silvicultural planning for large areas of reforested pine stands, optimizing their management through an optimal thinning schedule that maximizes net present value (NPV) and carbon (C) sequestration under wildfire risk. The specific objectives were to: (i) collect and process available forest inventory data to create an updated database using growth models and equations; (ii) estimate the C sequestered in existing stands and associate it with a wildfire risk value; and (iii) apply SP techniques to optimize thinning schedules, maximizing NPV and C capture while accounting for wildfire risk. The findings are discussed in the context of adaptive forest management, emphasizing the evaluation of forest carbon reserves and emission reduction programs.

3 Two-Stage Stochastic Optimization and Solution Using the Progressive Hedging Algorithm

Two-stage stochastic programming is a mathematical technique used to make optimal decisions under conditions of uncertainty. It is widely applied in situations where decisions must be made in two stages: (1) In the first stage, decisions are made before uncertainty is revealed. These decisions are proactive and are based on available data and probabilities of future outcomes. (2) In the second stage, decisions are made after uncertainty has been revealed. These decisions are adaptive and seek to mitigate risks or take advantage of opportunities based on observed outcomes.

The goal of two-stage stochastic programming is to minimize the expected cost (or maximize gains) across all possible scenarios, ensuring feasibility and consistency between the two stages. In this framework, the first-stage variables represent initial decisions made under uncertainty, while the second-stage variables correspond to scenario-specific decisions once the uncertainty has been resolved. Uncertainty is represented by random parameters (e.g., demand, prices, climate) and modeled through a finite set of scenarios, each with an associated probability. Constraints link the decisions of the first and second stages, ensuring feasibility across all scenarios.

The mathematical formulation of the two-stage stochastic optimization model can be defined as follows [1]:

Objective Function

$$min(x, x_s)\{c^T x + E_s[Q(x, \xi_s)]\}$$

where:

x: First-stage decision variables.

$c^T x$:: First-stage costs (deterministic).

$Q(x, \xi_s)$: Optimal second-stage costs, which depend on x and the realization of the random parameter ξ_s.

$E_s[Q(x, \xi_s)] = \sum_{s \in S} p_s Q(x, \xi_s)$: Expected second-stage cost across all scenarios.

where:

S: Set of all scenarios.

p_s: Probability of scenario s.

$Q(x, \xi_s) = \min(x_s)\{q_s^T x_s | T_s x + W_s x_s = h_s, x_s \geq 0\}$: Second-stage cost for scenario s.where:

x_s: Second-stage decision variables for scenario s.

$q_s^T x_s$: Second-stage costs (scenario-dependent).

T_s, W_s, h_s: Scenario-specific matrices/vectors linking the first and second stages.

ξ_s: Realization of the random parameter for scenario s.

First-stage constraints

$$Ax = 0, x \geq 0$$

where A represent deterministic parameters.

Second-stage constraints for each scenarios

$$T_s x + W_s x_s = h_s, x_s \geq 0$$

which links first-stage decisions to second-stage decisions for each scenario.

The Progressive Hedging (PH) method is a widely used iterative algorithm for solving two-stage and multi-stage stochastic programming problems. It was initially proposed by [17] as an efficient approach to decompose large, complex problems by dividing the scenarios into smaller subproblems that are solved independently.

In two-stage problems, the goal is to optimize decisions in the first stage (before uncertainty is revealed) and adjust these decisions in the second stage based on the outcomes observed in future scenarios. The PH method leverages this structure by dividing the problem into independent subproblems per scenario and using a penalty scheme to ensure consistency across scenarios.

The PH method has several advantages and limitations. Among its benefits, its ability to exploit the parallel structure of scenarios stands out, which accelerates computation in large-scale problems. Additionally, it is a flexible approach applicable to a wide range of two-stage and multi-stage problems. However, its convergence can be slow in cases where scenarios are highly correlated, and the algorithm's performance depends significantly on the proper selection of the penalty parameter (ρ). Despite these limitations, PH has established itself as an effective tool in applications such as natural resource planning, logistics, and finance, where handling multiple scenarios with high levels of uncertainty is required. The PH method relies on scenario-level decomposition and uses quadratic penalties to unify solutions from the subproblems in successive iterations.

The PH algorithm iteratively solves scenario subproblems while enforcing non-anticipativity constraints ($x_s = x \forall s \in S$, i.e., first-stage variables must be the same across scenarios) by introducing penalty terms. It gradually "hedges" toward a consensus solution by adjusting penalties and updating variables across iterations. PH commences with scenario decomposition, in which the problem is divided into a set of subproblems, each associated with an individual scenario, making them more manageable and enabling parallel resolution, significantly reducing computation time. In each iteration, a weighted average of first-stage decisions, called the aggregated or master solution, is calculated to ensure consistency of these decisions across subproblems (non-anticipative decisions). A quadratic penalty term is introduced to measure deviations of individual scenario decisions from the aggregated solution, ensuring convergence to a globally consistent decision. Iterations: Each subproblem is solved independently using the updated penalty, and the process of aggregating solutions and adjusting penalties is repeated until decisions converge to a consistent and feasible solution.

The steps of the algorithm are as [17]:

1. Initialization:

   a. Assign initial values to first-stage variables (x_s^0) for each scenario s.
   b. Set the penalty parameter $\rho > 0$, which determines the strength of the penalty for violating the non-anticipativity constraint.
   c. Initialize iteration counter k = 0.

2. Solve scenario subproblems (Decoupled):

   a. For each scenario s, solve the subproblem independently:

$$\min(x_s, y_s) f(x_S) + Q_s(x_s, \xi_s) + \rho \left|\left| x_s - \overline{x}^k \right|\right|^2$$

   b. The penalty term $\rho \left|\left| x_s - \overline{x}^k \right|\right|^2$ encourages x_s to align with the consensus variable $\overline{x}^k$

3. Update consensus variable ($\overline{x}^k$):

   a. Compute the average (consensus) of the first-stage decisions across scenarios:

$$\overline{x}^{k+1} = \frac{1}{|S|} \sum_{s \in S} x_s^{k+1}$$

4. Update penalty terms:

   a. Adjust the penalty terms (dual variables) for each scenario s:
$$\lambda_s^{k+1} = \lambda_s^k + \rho\left(\lambda_s^{k+1} - \overline{x}^{k+1}\right)$$

here λ_s^k is the dual variable associated with the non-anticipativity constraint.

5. Check convergence:

   a. Evaluate if the solutions across scenarios are sufficiently close (non-anticipativity constraint is approximately satisfied).
   b. The algorithm stops when:

$$\max(s \in S) \left|\left| x_s^{k+1} - \overline{x}^{k+1} \right|\right| \leq \epsilon$$

where ϵ is a small tolerance value.

6. Repeat:

   a. If convergence is not achieved, increment k and return to Step 2.

4 Case Study Description: Sierra De Filabres, Andalusia, Spain

The study area is located in Sierra de Filabres (Andalusia, Almeria, 37°22' N, 2°50' W; ca. 45,000 ha between 750 and 2168 masl; Fig. 1). The average annual rainfall is around 320 mm, while the monthly temperature ranges between 7 and 16 °C. The most commonly used species were pines, including *Pinus halepensis, P. pinaster, P. nigra and P. sylvestris*. Within Filabres, our study focused on forests owned by the Andalusian Forest Service, where most of the forest inventories were conducted (Fig. 1).

The forest inventory data for the study area were obtained from 2,302 inventory plots measured in 2007 for *P. halepensis, P. pinaster, P. nigra, and P. sylvestris*, respectively (radius = 12.6 m, slope correction applied). The data were updated to 2020 using equations from [5], [http://descargasrediam.cica.es/repo/s/RUR] (http://descargasrediam.cica.es/repo/s/RUR)). The forest inventory variables were generalized to the entire area using low-density LiDAR-ALS (Airborne Laser Scanning) models from the second coverage of the LiDAR-PNOA project, which was flown during the same year the inventory data were updated ([https://pnoa.ign.es/web/portal/pnoa-lidar/segunda-cobertura] (https://pnoa.ign.es/web/portal/pnoa-lidar/segunda-cobertura)).

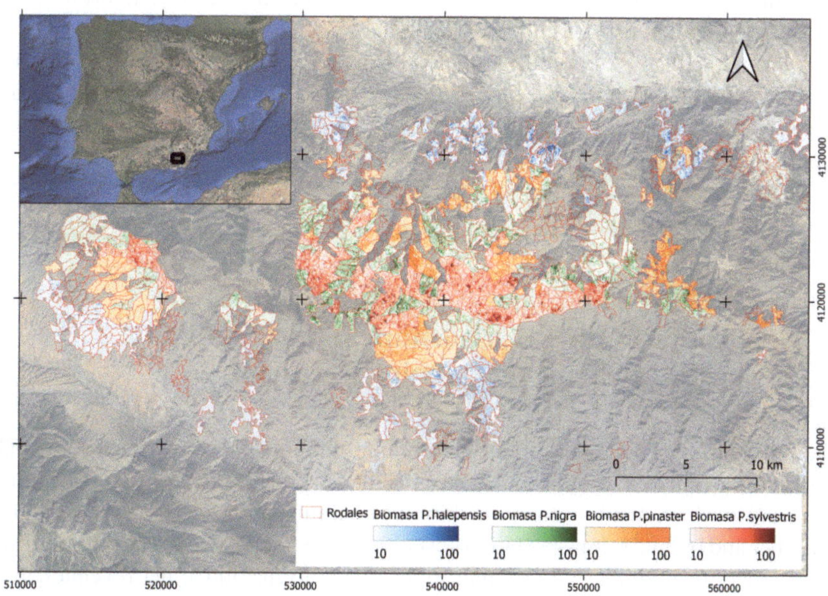

Fig. 1 Location of the study area in Sierra de Filabres (Almería, Spain)

Table 1 Dasometric characterization in 2020 of the pine plantations in Sierra de Filabres. Ho: dominant stand height in meters. N: number of trees per hectare in the stand. dq: quadratic mean diameter in centimeters. AB: stand basal area in m2 per hectare. Wt: stand biomass, summing all tree fractions

Variable	Statistic	P. halepensis	P. nigra	P. pinaster	P. sylvestris
Ho (m)	Mean	9.13	9.52	9.63	9.94
	StdDev	1.04	0.75	0.99	0.61
N (pies/ha)	Mean	528.12	599.43	571.78	645.11
	StdDev	74.12	81.16	86.30	61.13
d_q (cm)	Mean	17.17	17.92	18.29	18.96
	StdDev	2.53	1.76	2.23	1.33
AB (m^2/ha)	Mean	12.03	15.29	15.21	18.23
	StdDev	3.82	4.32	5.12	3.18
Wt (MgC/ha)	Mean	21.24	27.49	28.64	32.14
	StdDev	7.76	9.85	12.32	8.28

The methodology described in [12] was followed, which involved an initial filtering of point clouds to generate a surface model (2-m cell size). ALS metrics were then calculated for each inventory plot after normalizing the data by subtracting the Digital Terrain Model (DTM) from the point clouds. Finally, models for the forest inventory variables of interest were developed using the kNN algorithm, with distances computed using Random Forests (RF) [7].

All statistical analyses were conducted using R software version 4.2.1 [16]. The *usdm* package [11], version 2.1–6) was used for multicollinearity analysis, while variable selection and regression with kNN were executed using the *yaImpute* package, version 1.0–34 [2].

Table 1 presents a summary of the forest inventory variables calculated for the pine stands in Sierra de Filabres.

5 Application of the PH Algorithm in the Case Study: Stochastic Modeling of Thinning Decisions

The problem involves determining the optimal thinning decisions for a set of 630 stands to maximize economic and ecological benefits. This process entails deciding when and in which stands to carry out thinning operations, considering the availability of timber and biomass, transportation costs, and the potential occurrence of uncertain events such as wildfires. The primary objective is to maximize net revenues from the sale of timber, biomass, and carbon credits while minimizing operational costs and meeting constraints such as market demand (single sawmill for wood and single power plant for biomass) and the long-term sustainability of the forest.

The problem is modeled as a two-stage stochastic program. In the first stage, initial thinning decisions are made, which must be non-anticipative, meaning they must be valid across all uncertainty scenarios. In the second stage, scenario-specific adjustments are made. In this case, the second-stage decision includes determining how much timber and biomass to acquire externally to meet demand if internal production falls short due to events such as wildfires. For this analysis, five stochastic scenarios were used, each with a 20% probability of occurrence. These scenarios account for varying proportions of fire damage, ranging from 10% in Scenario 1 to 50% in Scenario 5.

The PH algorithm was used to solve this stochastic problem by combining scenario decisions in an iterative process aimed at finding a consistent solution. Through penalties and progressive adjustments, the algorithm ensures that the initial decisions are refined to balance costs and benefits across all scenarios. This approach facilitates adaptive forest management, optimizing economic outcomes while addressing the inherent uncertainties of forest systems.

Three thinning treatments were defined based on permanent plots from silvicultural treatments (Navarro [15, 13]: C (control or no thinning, 0% of the basal area), MT (moderate thinning: 30% of the basal area), and HT (heavy thinning: 50% of the basal area). The thinning intensity was assigned to each stand based on its current silvicultural condition and the management objectives outlined in the forest management plan, resulting in a reference silvicultural plan for each stand throughout the Special Plan.

The formulation of the thinning scheduling model under uncertainty to be solved using the PH algorithm is detailed below. Table 2 presents the set of sets, parameters, and variables used in the model.

5.1 Objective Function (oF)

$$\text{Maximize} \quad Z = Z_{first-stage} + Z_{second-stage} - Z_{penalty}$$

where,

$$Z_{\text{first - stage}}(X) = \sum_{i \in I} \sum_{t \in T} \Delta_t \cdot \mathrm{X}_{i,t} \cdot \mathrm{FCT}_i$$

Table 2 Sets, parameters, and decision variables of the stochastic model

Term	Definition
Sets	
I	Stands (630)
T	Periods (10)
S	Scenarios (5)
Parameters	
Δ_t	Discount factor in period t
p(S)	Probability of occurrence of scenario s
WPr	Price per ton of wood (sawlogs)
BPr	Price per ton of biomass (energy)
FCT_i	Fixed thinning cost for stand i
VTC	Variable thinning cost for stand i
CW	Cost per ton of wood (sawlogs) to be purchased from third parties
CB	Cost per ton of biomass (energy) to be purchased from third parties
ρ	Penalty in the PH algorithm
$\alpha_{i,s}$	Proportion of fire damage in compartment i under scenario s
Avail_ $W_{i,t}$	Availability of wood (sawlogs) in compartment i during period t
Avail_ $B_{i,t}$	Availability of biomass (energy) in compartment i during period t
DDW_i	Distance from stand i to sawmill
DDB_i	Distance from stand i to power plant (biomass)
TrC	Transport cost per t-km to sawmill or power plant
$DemW_t$	Demand of wood (sawlogs) in period t
$DemB_t$	Demand of biomass (energy) in period t
Variables	
$X_{i,t,s}$	Binary decision variable (first stage) that indicates whether stand i is harvested in period t under scenario s
$WD_{i,t,s}$	Wood (sawlogs) delivered from stand i, period t, and scenario s
$BD_{i,t,s}$	Biomass (energy) delivered from stand i, period t, and scenario s
$Flow\_W_t$	Flow of wood (sawlogs) from the stands to the sawmill in period t
$Flow\_B_t$	Flow of biomass (energy) from the stands to the power plant in period t
$ExtraW_t$	Additional wood (sawlogs) from third parties acquired in period t (second-stage decision variable)
$ExtraB_t$	Additional biomass (energy) from third parties acquired in period t (second-stage decision variable)
$\widehat{X_{i,t}}$	Consensus solution (average) of non-anticipative solutions

$$Z_{\text{second - stage}}(X, S)$$

$$= \sum_{s \in S} p(s) \cdot \sum_{t \in T} \Delta_t \left(\sum_{i \in I} \text{WD}_{i,t,s} \cdot \text{WPr} + \sum_{i \in I} \text{BD}_{i,t,s} \cdot \text{BPr} + \sum_{i \in I} \text{CStock}_{i,t,s} \cdot \text{PC} \right)$$

$$- \sum_{t \in T} \Delta_t \left(\begin{array}{c} \sum_{i \in I} (\text{WD}_{i,t,s} \cdot \text{VTC}) + \sum_{i \in I} (\text{WD}_{i,t,s} \cdot \text{TrC} \cdot \text{DDW}_i) \\ + \sum_{i \in I} (\text{BD}_{i,t,s} \cdot \text{TrC} \cdot \text{DDB}_i) + \text{ExtraW}_t \cdot \text{CW} + \text{ExtraB}_t \cdot \text{CB} \end{array} \right)$$

$$Z_{\text{penalty}} = \sum_{i \in I} \sum_{t \in T} \sum_{s \in S} \rho \left(\text{X}_{i,t,s} - \widehat{\text{X}_{i,t}} \right)^2$$

The first stage component of the OF includes the fixed thinning costs across stands and periods (discounted). The term captures planned thinning costs made before any scenario realization.

The second stage component of the OF includes the expected profit across scenarios: discounted revenue from wood/biomass and carbon stocks minus operations and procurement costs.

The penalty component of the OF penalizes inconsistent scenario-based decisions, encouraging non-anticipative solutions.

5.2 Constraints

Availability of timber and biomass in the stands: These constraints ensure that the amount of wood (WD) and biomass (BD) harvested from each stand i, in period t, under scenario s, does not exceed what is available, adjusted for fire damage.

$$\text{WD}_{i,t,s} \leq \text{X}_{i,t,s} \cdot \left(1 - \alpha_{i,s}\right) \cdot \text{Avail\_ W}_{i,t}$$

$$\text{BD}_{i,t,s} \leq \text{X}_{i,t,s} \cdot \left(1 - \alpha_{i,s}\right) \cdot \text{Avail\_ B}_{i,t}$$

Demand for timber and biomass. These constraints ensure that the market demand for wood (sawlogs) and biomass is met in every time period.

$$Flow\_W_t + ExtraW_t \geq DemW_t \quad \forall t \in T$$

$$Flow\_B_t + ExtraB_t \geq DemB_t \quad \forall t \in T$$

Flow of timber and biomass in each period. These constraints define the total flow of wood and biomass from stands to facilities in each period as the sum of what is harvested.

$$Flow\_W_t = \sum_{i \in I} WD_{i,t,s} \quad \forall t \in T$$

$$Flow\_B_t = \sum_{i \in I} BD_{i,t,s} \quad \forall t \in T$$

Each stand is thinned once over the Special Plan. This constrain ensures that each stand i can be harvested only once during the entire planning horizon, across all scenarios.

$$\sum_{t \in T} X_{i,t,s} = 1 \quad \forall i \in I, s \in S$$

Non-anticipativity constraint. This constraint ensures that the harvest decision for stand i in period t is the same across all scenarios (i.e., scenario-independent in the first stage).

$$X_{i,t,s} = \widehat{X_{i,t}}, \quad \forall i \in I, t \in T, s \in S$$

The two-stage stochastic problem solved using PH was implemented in Python, leveraging the open-source libraries *mpi-sppy* and *pyomo*, along with the Gurobi solver, which is available for free for academic use. The model was run on a computer equipped with an Intel® Core™ i9-13900HX processor (2.60 GHz, 24 cores) and 32 MB of RAM. For solving the subproblems within the PH algorithm, 4 cores were utilized.

The PH algorithm was set to run up to 50 iterations ($PHIterLimit = 50$) to achieve convergence. A quadratic penalty term ($defaultPHrho = 1.0$) controlled the degree of penalization for deviations from the aggregated solution, thereby ensuring convergence. In addition, a threshold of 1e-4 was specified ($convthresh = 1e\text{-}4$) to determine when the algorithm converged to a consistent and feasible solution. For parallelization, the PH algorithm was implemented with MPI (Message Passing Interface), in which each scenario subproblem was solved in parallel across multiple processes, significantly reducing computation time. The PH framework, provided by the *mpi-sppy* library, ensures that the master process aggregates results and adjusts penalties, while worker processes solve the individual subproblems.

In two-stage stochastic programming, several indicators or metrics are used to evaluate the performance of the solutions. These metrics allow for analyzing the value of considering uncertainty, the performance of deterministic solutions, and the quality of stochastic solutions [1]. Table 3 presents the key metrics, their definitions or calculation formulas, and a description of their significance.

Table 3 Key metrics, components, and description of a stochastic model

Metric	Definition or formula	Description
Stochastic Solution Value (STO)	Optimal value of the stochastic programming model	Represents the best value considering all scenarios and probabilities. Reflects the value of solving the problem using PH
Expected Value Problem (EVP)	Optimal value of the deterministic problem (using expected values)	Solution obtained by solving the problem deterministically, replacing random variables with their expected values
Expected Value with Perfect Information (EVSS)	$EVSS = \sum_{sCS} p_s z_s^*$ where z_s^* is the optimal value for scenario s	Represents the best outcome for each individual scenario assuming perfect information
Expected Value of Perfect Information (EVPI)	EVPI = EVSS—STO	Measures the value of knowing each scenario in advance (benefit of perfect prediction)
Expected Value of the Deterministic Solution (EVWS)	Objective value when applying the deterministic solution to all scenarios	Represents the expected value of the deterministic solution (from EVP) when applied to all scenarios with their probabilities
Value of the Stochastic Solution (VSS)	VSS = STO–EVWS	Measures the improvement resulting from using the stochastic solution instead of the deterministic one
Relative EVPI	Relative EVPI = (EVPI/STO) × 100%	Proportion of the stochastic solution value attributed to having perfect information
Relative VSS	Relative VSS = (VSS/STO) × 100%	Proportion of the stochastic solution value attributed to including uncertainty in the problem

6 Results

6.1 Solutions for the Two-Stage Stochastic Model

The metrics obtained from the solutions of the deterministic and stochastic models are presented in Table 4. The high VSS value (€284,439) demonstrates that solving the problem stochastically provides a clear advantage over the deterministic approach. A relative VSS of 2.55% highlights the improvement achieved by considering uncertainty, even if the problem's sensitivity to such uncertainty is not extreme. In contrast, the EVPI is very low (€4,340), indicating that perfect information does not significantly improve the outcomes. This suggests that the current stochastic model effectively captures variability. A relative EVPI of 0.04% emphasizes that perfect information has minimal added value.

Table 4 Metrics obtained from the solutions of the deterministic and two-stage stochastic models

Metric	Value (€)
STO	11,174,400
EVP	11,189,629
EVSS	11,178,739
EVPI	4,340
EVWS	10,889,961
VSS	284,439
Relative EVPI	0.04%
Relative VSS	2.55%

On the other hand, the deterministic solution performs poorly when tested across all scenarios (€10,889,961), confirming that ignoring uncertainty leads to suboptimal results. This gap highlights the importance of using a stochastic approach in problems with significant variability between scenarios. The STO (€11,174,400) is slightly lower than the EVSS (€11,178,739), indicating that while stochastic optimization performs well, there is a small theoretical improvement possible under perfect information. The EVP (€11,189,629) is only slightly higher than the STO, illustrating the excessive optimism of the deterministic approach when ignoring scenario variability. This analysis confirms that stochastic optimization effectively balances uncertainty and performance, providing a practical and efficient framework for solving problems like thinning planning.

6.2 *Thinning Scheduling*

The number of stands thinned and the total area intervened throughout the Special Plan when applying the two-stage stochastic optimization model are shown in Fig. 2. The number of stands thinned is high at the beginning (93 stands in year 1) and then decreases to a minimum of 11 in year 6. After this point, there is a steady increase, reaching the highest value of 139 stands in year 10. This trend suggests a planned reduction in thinning operations during the intermediate periods to optimize growth toward the end of the planning period.

The total area subjected to thinning interventions follows a trend similar to the number of stands, starting with a high value (4,489 ha in year 1), decreasing to a minimum of 745 ha (year 6), and then increasing sharply to a peak of 7,442 ha (year 10). The close relationship between the number of stands and the thinning area indicates a well-distributed thinning approach across the stands, with no significant variations in the average stand size between periods. For example, in year 1, the optimization function schedules thinning in 93 stands with an average area of ~ 69.7 ha per stand, whereas in year 10, 139 stands are scheduled for thinning with an average of ~ 53.6 ha per stand. Although the number of stands increases toward the

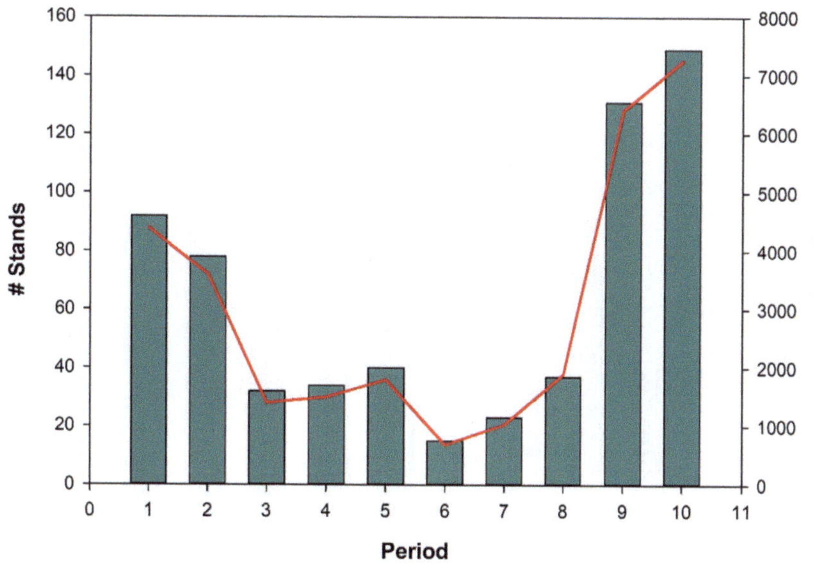

Fig. 2 Number of thinned stands and thinning area per period according to the stochastic solution

later periods, the average area per stand seems to decrease slightly, which could reflect smaller stands being thinned at the end of the Special Plan or improved efficiency in selecting thinning areas.

6.3 Additional Timber from Third Parties to Meet Demand

Figure 3 illustrates the optimal timber volume (expressed as biomass and carbon) extracted per period. The volume increases consistently until year 8. However, in years 9 and 10, there is a noticeable decline in the extracted volume across all scenarios associated with fire risk. Furthermore, as fire risk intensifies, there is a general trend of increased timber extraction within the thinning program. This trend is evident in Fig. 3, where the timber volume from thinning rises proportionally with the fire damage proportion (10%, 20%, 30%, 40%, and 50%). As the fire damage proportion increases from 10 to 50%, the timber volume required under the thinning plan nearly doubles, highlighting the strong dependence on thinning as a strategy to mitigate the impacts of significantly heightened fire risk.

Considering the results from Figs. 2 and 3, some relationships between the two can be identified. In Fig. 2, the years where the plan recommends intervening in a greater number of stands and larger areas are years 1–2 and 9–10. This leads to an increase in timber supply between years 1 and 2, which remains relatively stable until year 9, before dropping sharply in year 10 (as shown in Fig. 3). Additionally, the results in Fig. 3 show that the volume varies depending on fire risk. However, in years

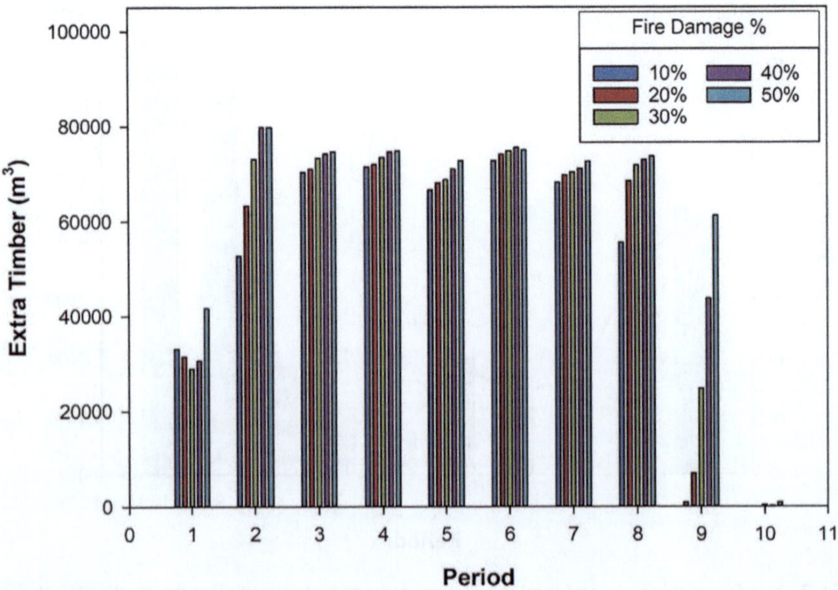

Fig. 3 Wood volume production throughout the Special Plan (Years 1 to 10) according to the stochastic solution

1 and 10, this effect is more pronounced, which could be explained by the increase in the number of stands and the area subject to thinning during these two periods (Fig. 2). In the intermediate years (years 3 to 8), both figures show a notable decline in thinning interventions, but the extracted volume remains relatively stable. This suggests a management strategy that prioritizes volume optimization during these periods while reducing the number of thinned stands (e.g., to minimize harvesting costs).

7 Discussion

7.1 Importance of Stochastic Models for the Thinning Problem Under Fire Risk

Two-stage stochastic programming, based on the estimation of the Expected Value of Perfect Information (EVPI) and the Value of the Stochastic Solution (VSS), provides valuable insights for silviculture planning in artificial pine stands under fire risk by accounting for the uncertainty associated with such events.

Our results show a low EVPI, resulting from the small difference observed between the stochastic solution and the perfect information solution. This suggests that knowing the uncertainty in advance (perfect information) does not significantly improve the solution's performance. This can be explained by the type of forest system considered (monospecific and regular artificial stands with low structural and biomass variability), which makes their response to thinning very robust against fire-related uncertainties. This means that the optimal stochastic solution performs well across different scenarios. It also indicates that the impact of uncertainty on the objective function is minimal, which could be attributed to the limited variability or low sensitivity to uncertain parameters. In our case, each scenario has an equal probability of occurrence (20%), and the proportion of fire damage varies between 10 and 50%. Higher EVPI values would be expected in more complex situations that consider a larger number of scenarios characterized by greater variability in fire risk and the likelihood of such damage.

In contrast, a relatively high VSS value was obtained, reflecting the significant advantage of the stochastic solution compared to the deterministic solution based on the Expected Value Solution (EVS). This indicates that considering the uncertainty associated with fire risk in the decision-making process leads to a much better solution than relying on deterministic approximations. The deterministic solution (EVS), even in a reduced-scenario case like our study, may perform poorly under uncertainty as it optimizes only for the average scenario, ignoring variability. This becomes even more relevant in more complex situations with a high number of scenarios characterized by significant variability, making it essential to include uncertainty in the forest management decision-making process for such systems. In this way, stochastic optimization better captures the trade-offs specific to each scenario, resulting in a more adaptable and effective solution.

The combined effect of a low EVPI and a high VSS implies that, while knowing the exact fire risk would not significantly improve the model's performance, accounting for the uncertainty associated with this risk greatly enhances the solutions. This typically occurs in problems where the stochastic solution is inherently robust across different scenarios, but the deterministic solution (EVS) fails to effectively account for variability. This combination underscores the importance of stochastic programming in capturing the nuances of uncertainty without over-relying on perfect predictions.

7.2 *Impact of Fire Risk and Uncertainty on Thinnings*

The results obtained regarding thinning schedules provide a complementary perspective on forest planning under uncertainty, such as fire risk, and its impact on stand-level management associated with the need for external resources. Our findings show that the number of thinned stands and the total area subject to thinning throughout the Special Plan exhibit similar patterns: high levels of intervention in the initial and final years, with a significant decline during the intermediate periods, reaching a minimum in year 6. This behavior could be linked to a silvicultural strategy aimed at minimizing fire risk at the beginning and maximizing production toward the end of the Special Plan. During the intermediate periods, silviculture seeks to maintain a consistent volume output while reducing fire risk.

Additionally, it is evident that the increase in timber extraction grows proportionally with fire risk, peaking in year 8. However, this need decreases toward the end of the Special Plan (years 9 and 10), which may be related to the increase in thinning activities during that period, as shown in Fig. 2. This indicates that the thinning plan offsets timber production even under scenarios of high uncertainty due to fire risk.

The relationship between the number of stands to be thinned and the volume extracted highlights the importance of an integrated approach to forest planning. The synchronization between years with high thinning activity and timber supply reflects a strategy aimed at balancing forest planning with the reliance on external resources. However, the peaks in extracted volume during year 10 suggest that there is room to optimize operations and reduce vulnerability in scenarios of severe damage.

Overall, these results underscore the need to adopt silvicultural planning strategies that incorporate both fire risk and the optimization of biomass production as an adaptation strategy (e.g., carbon sequestration capacity) in the face of increasing uncertainty associated with heightened fire risk. A planning approach that combines efficient silvicultural practices with external demands for environmental services (e.g., carbon markets) will help ensure the resilience of forest systems in the face of future challenges.

7.3 *Thinning Actions and Impact on Fire Risk*

In this study, each stand is assumed to undergo only one thinning action within the 10-year planning horizon. This reflects the standard forest management practices followed by the forest administration in Andalucía, Spain, where thinning is typically scheduled once per rotation due to ecological, logistical, and administrative considerations. Although the model does not currently allow for multiple or variable thinning intensities per stand, this simplification ensures consistency with real-world planning while maintaining computational tractability. Nonetheless, future versions of the model could be extended to incorporate more flexible thinning regimes to better support adaptive management strategies.

Additionally, while thinning is known to influence both the likelihood and severity of wildfires, this feedback is not explicitly modeled here. Instead, fire risk is treated as an exogenous factor: each scenario is defined by a fixed proportion of fire damage that does not depend on management actions. The observed increase in volumes extracted under higher fire risk scenarios should thus be interpreted as a response to greater expected losses of timber, rather than as a preventive strategy reducing fire risk. We acknowledge this modeling limitation and understand that incorporating fire-thinning interactions would enhance the realism of the framework. Future research will address this by coupling the current stochastic optimization model with a spatially explicit fire behavior model or by allowing fire risk to be endogenously adjusted based on stand conditions and management interventions.

8 Conclusions

The results of this study underscore the importance of stochastic programming as a tool to manage the inherent uncertainty of fire risks in thinning planning. The low EVPI indicates that precise knowledge of future damages would not significantly improve the solution, reflecting the robustness of the stochastic solution across diverse scenarios. This suggests that the impact of uncertainty on the objective function is relatively limited in our case study, though different EVPI values could emerge in more complex problems with greater variability between scenarios. On the other hand, the high VSS demonstrates that incorporating uncertainty into decision-making results in far more effective solutions compared to deterministic approaches. This becomes particularly relevant in situations with higher variability, where stochastic optimization better captures the specific trade-offs of each scenario.

The relationship between the observed patterns in the number of stands thinned, the area intervened, and biomass production highlights the importance of an integrated approach to forest planning. The increase in volume during the intermediate and final years of the Special Plan, particularly in years 8 and 9, followed by a decrease in year 10, underscores the significance of synchronizing internal management operations with external needs. Together, these findings suggest that adaptive forest planning, which incorporates strategies to minimize uncertainty and efficiently respond to external demands, is essential to ensure the sustainability and resilience of the forest system in the face of future climatic and economic challenges.

Acknowledgements This study was funded by the OptiForValue project—Optimized forest-based value chains for high-value applications & improved forest management (HORIZON-CBE-JU-2023, Project ID: 101157658). Additionally, it received funding from DesFutur, which has the support of the Biodiversity Foundation of the Ministry for the Ecological Transition and the Demographic Challenge (MITECO) of the Government of Spain, within the framework of the Recovery, Transformation and Resilience Plan (PRTR), funded by the European Union—NextGenerationEU. We also acknowledge the support of the Ministry of Sustainability, Environment, and Blue Economy (Junta de Andalucía).

References

1. Birge, J.R., Louveaux, F.: Introduction to Stochastic Programming, 2nd edn. Springer, New York, NY (2011)
2. Crookston, N.L., Finley, A.O.: YaImpute: an R package for k-NN imputation. J. Stat. Softw. **23**(10), 1–16 (2007)
3. De Pellegrin Llorente, I., Eyvindson, K., Mazziotta, A., Lämås, T., Eggers, J., Öhman, K.: Perceptions of uncertainty in forest planning: Contrasting Forest professionals' perspectives with the latest research. Can J For Res, **53**(6), 391–406 (2023). https://doi.org/10.1139/cjfr-2022-0193
4. De Rigo, D., Libertà, G., Durrant, T.H., Vivancos, T.A., Sanmiguel-Ayanz, J.: Forest fire danger extremes in Europe under climate change: variability and uncertainty. Doctoral Dissertation, Publications Office of the European Union (2017)
5. Guzmán Álvarez, J.R., Venegas Troncoso, J., Seseña Rengel, A.: Modelos de crecimiento y producción de biomasa de los pinares de Andalucía. 6CFE01–033. Actas 6 Congreso Forestal Español (2013)

6. Kangas, A., Kurttila, M., Hujala, T., Eyvindson, K., Kangas, J.: Decision support for forest management 2nd ed., Manag For Ecosyst, **30**(307) (2015). https://doi.org/10.1007/978-3-319-23522-6
7. Liu, Z., Peng, C., Work, T., Candau, J.N., Desrocherss, A., Kneeshaw, D.: Application of machine-learning methods in forest ecology: recent progress and future challenges. Environ. Rev. **26**(4), 339–350 (2018)
8. Lorenz, K., Rattan Lal, Y.: Carbon Sequestration in Forest Ecosystems. Springer Springer Dordrecht (2010)
9. Montero, G., Ruiz-Peinado, R., Munoz, M.: Producción de biomasa y fijación de CO2 por los bosques españoles. Ministerio de Medio Ambiente, Madrid (2011)
10. Moreno, J.M., Morales-Molino, C., Torres, I., Arianoutsou, M.: Fire in Mediterranean pine forests: past, present and future. In: Pines and Their Mixed Forest Ecosystems in the Mediterranean Basin, pp. 421–456 (2021)
11. Naimi, B, Hamm, N.A., Groen, T.A., Skidmore, A.K., Toxopeus, A.G.: Where is positional uncertainty a problem for species distribution modelling. Ecography **37**, 191–203 (2014)
12. Navarrete-Poyatos, M.A., Navarro-Cerrillo, R.M., Lara-Gómez, M.A., Duque-Lazo, J., Varo, M.D.L.A., Palacios Rodriguez, G.: Assessment of the carbon stock in pine plantations in Southern Spain through ALS data and K-nearest neighbor algorithm based models. Geosciences **9**(10), 442 (2019)
13. Navarro Cerrillo, R.M., Cachinero-Vivar, A.M., Pérez Priego, Ó., Cantón, R.A., Begueria, S., Camarero, J.J.: Developing alternatives to adaptive silviculture: thinning and tree growth resistance to drought in a Pinus species on an elevated gradient in Southern Spain. Forest Ecol. Manag **537**, 120936 (2023)
14. Navarro-Cerrillo, R.M., Duque-Lazo, J., Rodríguez-Vallejo, C., Varo-Martínez, M.Á., Palacios-Rodríguez, G.: Airborne laser scanning cartography of on-site carbon stocks as a basis for the silviculture of Pinus halepensis plantations. Remote Sens. **10**(10), 1660 (2018)
15. Navarro Cerrillo, R.M., Sánchez Salguero, R., Rodríguez, C., Lazo, J.D., Moreno Rojas, J.M., Palacios Rodríguez, G., Camarero, J.J.: Is thinning an alternative when trees could die in response to drought? The case of planted Pinus nigra and P. Sylvestris stands in southern Spain. Forest Ecol. Manag. **433**, 313–324 (2019)
16. R Core Team.: R: a language and environment for statistical computing. R Foundation for Statistical Computing, Vienna, Austria. URL https://www.R-project.org/. (2022)
17. Rockafellar, R.T., Wets, R.J.B.: Scenarios and policy aggregation in optimization under uncertainty. Math. Oper. Res. **16**(1), 119–147 (1991)
18. Rodrígez Silva, F.R., Martínez, J.R., González-Cabán, A.: A methodology for determining operational priorities for prevention and suppression of wildland fires. Int. J. Wildland Fire **23**(4), 544–554 (2014)
19. Ruiz-Peinado, R., Oviedo, J.A., Senespleda, E.L., Oviedo, F.B., Del Río Gaztelurrutia, M.: Forest management and carbon sequestration in the Mediterranean region: a review. Forest Syst. **26**(2), 10 (2017)
20. Shapiro, A., Dentcheva, D., Ruszczyński, A.: Lectures on Stochastic Programming: Modeling and Theory (2nd ed.). Philadelphia, PA: Society for Industrial and Applied Mathematics (SIAM) (2009)
21. Tedim, F., Leone, V., Xanthopoulos, G.: Wildfire risk management in Europe: the challenge of seeing the "forest" and not just the "trees." In: Proceedings of the 13th International Wildland Fire Safety Summit & 4th Human Dimensions of Wildland Fire, Managing Fire, Understanding Ourselves: Human Dimensions in Safety and Wildland Fire, pp. 213–238 (2015)
22. Wallace, S.W., Ziemba, W.T. (eds.): Applications of Stochastic Programming. Springer, New York, NY (2005)
23. Wei, Y., Rideout, D., Kirsch, A.: An optimization model for locating fuel treatments across a landscape to reduce expected fire losses. Can J For Res, **38**(4), 868–877 (2008). https://doi.org/10.1139/X07-162
24. Xu, W., Shrestha, A., Wang, G., Wang, T.: Site-based climate-smart tree species selection for forestation under climate change. Clim Smart Agric, **1**(2), 100019 (2024) https://doi.org/10.1016/j.csag.2024.100019

Advances in Decision Support Platforms for Prioritizing Investments in Forest and Rangeland Restoration, Risk Reduction and Biodiversity Conservation

Alan Ager and Hugh Safford

Abstract The growing interest in spatial prioritization models and their deployment to support forest restoration and conservation efforts is substantiated by the emergence of many new landscape planning platforms. In this paper we review selected landscape planning and prioritization platforms and their application in diverse ecosystems and operational environments. We define platforms as integrated solutions that leverage one or more technology building blocks including big data, spatial optimization algorithms, geospatial technology, and cloud computing. We pay particular attention to contrasting platforms developed to prioritize active forest management as part of restoration and risk reduction efforts on western US landscapes, versus other systems that were developed to identify efficient conservation reserve designs to protect biodiversity and carbon in temperate and tropical ecosystems. The models examined represent an array of planning systems with a common thread in all the platforms of the use of spatial optimization to find efficient solutions in terms of minimizing the cost per unit of progress towards stated targets. Divergence in platform designs, data usage, scale, resolution, and intended users was expected given the diversity of spatial planning problems for which they were developed. Highlighting the design differences among these platforms provided insights on the contribution of specific features to their functionality. One major finding was that the spatial planning problem and its formulation diverge significantly between widely used "top-down" conservation platforms and "bottom-up" tools in current use to optimize investments in active forest management as part of wildfire risk reduction and restoration programs. We conclude that the uptake of these newer platforms and underlying science by public and private entities is not keeping pace with technology, and thus the bottleneck in application and realizing potential benefits rests with institutional capacity to absorb and implement new technologies rather than the design, development, and deployment of them.

A. Ager (✉)
Oregon State University, Corvallis, OR, USA
e-mail: alan.ager@oregonstate.edu

H. Safford
Department of Environmental Science and Policy, University of California, Davis, CA, USA

Vibrant Planet, Truckee, CA, USA

F. Alvelos et al. (eds.), *Advances in Optimization and Wildfire*, Lecture Notes in Operations Research, https://doi.org/10.1007/978-3-032-03108-2_2

Keywords Spatial planning · Forest planning · Decision support · Biodiversity conservation · Restoration planning · Landscape planning

1 Introduction

Prioritizing investments in restoration, risk reduction, and conservation programs is an important part of the planning process to solve a wide range of resource management issues in the US and elsewhere [17, 22, 36, 43, 48, 63]. Newer planning and prioritization platforms are rapidly emerging from academic institutions, the public sector, and non-government entities to meet conservation, restoration and risk mitigation initiatives in diverse ecological systems ([20, 34, 41, 49, 56, 69, 74], Safford in review). Much of the development of these platforms has been motivated by the widening global backlog in restoration needs and growing threats to biodiversity from natural and anthropogenic hazards [70].

Newer platforms potentially bridge landscape planning with the science of scenario analysis that emphasizes the importance of examining large arrays of management actions and considering uncertain climate and disturbance [45]. Scenario analysis is an underutilized tool, especially in federal land management [64] despite its proven effectiveness in planning [10, 60]. Newer platforms leverage multiple technology advances in operations research including cloud computing, big data, artificial intelligence (AI), simulation modeling, informatics, geospatial science, machine learning, and risk science, and potentially create novel solutions to address many of the planning problems that face land managers and policymakers globally [14]. Although newer platforms have been developed for application in many ecosystems, a common functionality is the use of spatial optimization algorithms to find cost-effective solutions to conservation and restoration problems [3, 11, 17, 36, 39, 48, 63], Safford in review). Spatial optimization is a keystone concept in landscape planning, where mathematical and computational techniques are employed to build efficient plans in terms of cost and understand tradeoffs. Analytical foundations for spatial optimization include various forms of mathematical programming (linear, integer, mixed integer), and heuristic approaches like greedy search, Tabu and various genetic algorithms [8].

Although there is a diverse array of conservation and restoration planning problems that that can benefit from spatial optimization and supporting technologies, one in particular is ecological restoration and wildfire risk reduction in the western US, where huge tracts of public lands are increasingly impacted by ecologically and socio-economically destructive wildfire [67]. Federal and state wildfire risk reduction and ecological restoration programs require prioritization to address a large backlog of forest management needs under significant budget and capacity constraints [3]. Data to support prioritization are widely available in the numerous Federal, State, and local assessments of landscapes conditions and restoration needs, including ecological departure, watershed health, wildlife habitat, and terrestrial conditions [18, 42, 65]. Rapid expansion of risk estimation technologies and wildland fuels mapping has led to proliferation of risk and exposure assessments and a number of prioritization systems have been developed over the past two decades [52, 53] although none in operational use have incorporated spatial optimization technologies that are increasingly used in systematic conservation planning [63, 74]. Clearly, there is a growing need to translate land condition and natural hazard assessments into spatial management plans that prioritize investments to maximize outcomes to address the wildfire crisis in the US [67] and in other fire prone regions of the globe.

Given the emergence of many new prioritization and optimization platforms, a synthesis of these system is warranted to understand how newer technologies have been brought to bear on specific planning problems, and to assess the relative strengths and weaknesses of the various approaches. In this paper, we examined a selected set of newer platforms to understand both parallel functionality and divergent technological solutions to the overall problem of planning and prioritization for ecological restoration, biodiversity conservation, and wildfire risk reduction. We restricted the review to platforms that integrated multiple newer technologies to apply spatial optimization methods to directly serve a broad community of practitioners rather than develop technical approaches to solve specific spatial planning problems [8]. Specific evaluation criteria included descriptive websites, open access to the software or outputs, use of spatial optimization, spatially explicit representation of landscapes, cloud deployment, and supporting publications with applications on real-world problems. We focused on platforms that employed multiple newer technologies, and integrated spatial optimization to simulate management scenarios, optimize priorities, analyze tradeoffs, and predict outcomes. Following our comparative review, we describe a case study application of one platform, the Vibrant Planet Decision Support System (Safford in review), which is a commercial product that leverages multiple new technologies to streamline the planning on US public forests. We then discuss the overall science findings, management implications, and future development with external technology partners as part of broader application of the model to optimize the implementation of landscape restoration, conservation, and natural hazard risk reduction.

2 Review of Selected Planning Platforms

We selected 8 specific platforms based on the criteria above and used information from both publications and websites to obtain information needed for a comparative assessment of technologies and their application. The platforms examined here were Marxan, PrioritizR, Weplan-Forests, Plangea, PReSET, Forsys, Vibrant Planet's Decision Support System (VPDSS), and Planscape (Table 1). These platforms represent a fraction of the published spatial planning models as reported in a recent systematic review [see 24] but we note that the vast majority of published models were developed to solve specific research problems rather than meet the broader need for operational platforms. Our sample of platforms was sufficient to highlight the diversity of approaches and contexts for the application of spatial optimization methods. We note that some of the descriptions are based on incomplete online descriptions. We specify where technical aspects were not entirely clear from the available materials.

Table 1 Key information on spatial planning platforms described in this review. Information was obtained from published literature and available web materials

Platform	Organization	Optimization method	Demonstrated scale of application (ha)	Website	Key scientific publications and manuals
PReSET	USGS, Fort Collins Science Center	Integer programming	Local to > 20 million	https://www.usgs.gov/centers/fort-collins-science-center/science/prioritizing-restoration-sagebrush-ecosystems-tool	[22]
ForSys	ForSys Research Consortium	Greedy heuristic	5000 to > 100 million	https://www.forsysplanning.org/	[20]
Planscape	The Planscape Cooperative	Patchmax Greedy heuristic [25]	Local to millions	https://www.planscape.org/	https://www.planscape.org/documentation/user-guide/
VPDSS	Vibrant Planet	Patchmax greedy heuristic [25]	3000 to > 5 million	https://www.vibrantplanet.net	[68], Safford in review
Plangea	International Institute for Sustainability	Linear programming	Up to 3 billion ha	https://www.iis-rio.org/en/projects/plangea/	[63]
PriortizR	Research team	Mixed-integer linear programming	46 million	https://prioritizr.net/	Hanson et al. 2023
Marxan	Marxan.org	Simulated annealing	Local to > 35 million ha	https://marxansolutions.org/	[71]
WePlan-Forests	International Institute for Sustainability	Linear programming	10 million to 800 million ha	https://www.weplan-forests.org/	WePlan-Forests

2.1 Marxan Planning Platform

Marxan is a decision-support tool originally developed for systematic conservation planning and spatial prioritization of conservation areas. Marxan is the most widely used spatial planning platform that employs optimization to find efficient solutions to conservation and restoration problems [7]. The literature on Marxan case studies is extensive, including multiple review papers [24, 35, 37, 71], and the reader is pointed to these and other recent publications for more details. Marxan was originally designed for marine reserve systems based on the terrestrial reserve design software Spexan [6]. In many applications Marxan has been used to select potential planning areas to optimize ecological, social, and economic criteria while considering connectivity of the selected patches as measured by boundary lengths between selected and non-selected planning areas.

The original Marxan used simulated annealing to optimize conservation networks to achieve habitat conservation goals while minimizing costs. In a typical application it ingests data on species distributions and other factors to create optimized conservation reserves. Recent studies [29]; and see Prioritizr description below) have examined the performance of the Marxan optimization algorithm (simulated annealing) versus linear programming and these have reported significantly faster linear programming solutions with exact optimality. Most recently Marxan development was expanded with a partnership between The Nature Conservancy, Microsoft, the European Commission, and the Marxan Community of Practice and a new cloud version was released with improvements to the design to further mainstream spatial decision support systems for prioritizing investments in conservation (https://marxan solutions.org/marxanmapp/).

Marxan is a relatively complex system to parameterize and calibrate compared to several of the other platforms discussed in this paper largely because the model has a wide range of options to formulate and solve conservation planning problems. Like other platforms, planning problems are formulated with targets, goals, spatial planning units, and a wide array of options to specify costs and benefits. The term "Target" is used to describe the amount of each goal for a feasible solution. In general, Marxan solutions minimize an objective function that includes both the cost of the planning unit and its boundary length, with the latter variable favoring solutions that are spatially compact. The boundary penalty can be varied by the user to achieve the desired density of planning units within a larger spatial domain. The choice of the boundary length modifier is subjective and achieving the desired result is an interactive process and dependent on the planning problem at hand. For a given reserve system, a lower boundary length results in a more compact and less fragmented solution. A randomization feature also introduces some variability in the solutions in terms of the composition of planning areas, and a second report is generated that specifies the number of times each planning area was in the best solution.

2.2 PrioritizR

PrioritizR is an R package for systematic conservation that shares many features with Marxan and is a close relative in terms of functionality and purpose, with the major divergence being that the platform was developed in R [29] and uses external mixed integer linear programming (MILP) for optimizing priorities. In a recent application the platform was used to prioritize conifer removal in the sagebrush (*Artemisia* spp.) steppe ecosystem in western North America to improve habitat [51]. The platform description online (https://prioritizr.net/) asserts that PrioritizR finds cheaper solutions in a shorter period of time than Marxan using linear programming methods. The prioritizR R package can call multiple linear programming solvers [29] although the developers recommend the Gurobi [28] software suite to solve a problem formulation. Typical scenarios use cost data associated with converting or managing planning units for conservation in the objective functions. According to the developers, the cost values associated with land use conversions or land purchase can be represented with surrogate data that describe, for instance, urban, density or loss of commercial value. In addition to costs and targets, scenarios require planning (decision) units, features, and management zones along with an objective, which is typically related to the cost to conserve planning units. Global penalties on fragmentation of the selected planning areas are used to control the level of aggregation in the solutions. Targets specify the desired quantity of a feature required in a solution and constraints are used to specify required conditions. Marxan-like portfolios can be generated that build solutions using randomization to build hundreds of solutions that are within some level of the optimum solution. Testing of the system by the authors with knowledge of R found it relatively easy to take a spatial dataset created for other platforms like ForSys and VPDSS and use it in PrioritizR to analyze scenarios. Although PrioritizR does not provide an integrated platform (e.g. geospatial interface, cloud hosting, pre-packaged data) the relative ease that users can formulate and execute spatial optimization problems using the open-source R package merited its inclusion in this review.

2.3 WePlan—Forests

WePlan-Forests is a decision support platform designed for restoration planning in tropical and subtropical forest ecosystems in developing countries [31]. The system was developed by the International Institute for Sustainability Australia and International Institute for Sustainability Rio to identify national scale restoration goals that maximize the benefits of climate change mitigation, biodiversity conservation, and poverty alleviation. The model was built for application in subtropical forests within ± 25 degrees of the equator and is applied at national scales with a 1 km^2 resolution. The WePlan-Forests website provides reports detailed optimization results for 37 nations with detailed publications for selected areas [72]. The platform is a spatially explicit, forest restoration planning tool that uses linear programming to optimize and evaluate alternative scenarios for restoration at the national scale. The solutions report the benefits, costs and spatial distribution of national restoration priorities for each nation. The intended use is to evaluate large numbers of scenarios that vary in targets to provide decision-makers with estimates of the climate mitigation and biodiversity benefits that may accrue from various levels of forest restoration, and the opportunity and establishment costs associated with that restoration. These analyses can be used to address specific questions related to scenarios that maximize returns-on-investments, trade-off between carbon and biodiversity. The developers recognize that WePlan-Forests solutions are not site specific in terms of specific restoration activities, since local factors including governance and land ownership can play a role in deciding restoration investments and relative priorities. WePlan-Forests uses data generated by Conservation International as part of the SPARC Project [58] that include data include species occurrence records, expert-verified species ranges, species' traits and phylogenies, land cover, protected areas, and climate surfaces. Using a combination of statistical and process-based models, SPARC has produced species range maps for more than 100,000 tropical plant and animal species, under current climate and a variety of projected future climate scenarios. The specific response metrics in the model include the change in carbon sequestration per unit of restoration as a measure of climate change mitigation benefit, and biodiversity conservation benefit estimated as the average reduction in extinction risk for all forest-associated species. The model uses constraints on implementation costs of forest restoration and considers the potential cost savings from natural regeneration. The platform analyzes trade-offs and production frontiers between climate change mitigation and biodiversity conservation benefits and allows the user to explore compromise solutions. Scenarios are run by the development team with constraints to limit restoration to five pre-determined levels, ranging from 10 to 50 percent of the area that was determined as available for restoration. WePlan-Forests is not directly executed by end users, rather, scenarios are formulated and executed by the development team [31] and the results are posted on a server and accessed with an interface allowing end users without expertise in spatial modeling and optimization explore the results. This centralized processing model has many advantages in modelling environments where technology and expertise for relatively complicated mathematical programming models are not widely available. Results are accessed by the user using the SEPAL (System for Earth Observation Data Access, Processing and Analysis for Land Monitoring), which allows users to retrieve and manipulate the geospatial data resulting from WePlan (https://openknowledge.fao.org/handle/20.500.14283/cc1803en). Developed by the Food and Agriculture Organization of the United Nations SEPAL was designed to assist national forest monitoring and reporting for the Reducing Emissions from Deforestation and Forest Degradation, Forest Conservation, Sustainable Management of Forests and Enhancement of Carbon Stocks in Developing Countries (REDD +) mechanism. An extensive library of reports has been developed for 37 countries listed on the WePlan website.

2.4 PLANGEA

PLANGEA is a similar system to WePlan-Forests and was apparently the progenitor although this is not explicitly stated in available documentation. PLANGEA was developed by the International Institute for Sustainability (IIS) to optimize the spatial planning of ecosystem restoration and conservation at large, national scales. The model was intended to help develop national strategies to implement the new Kunming-Montreal Global Biodiversity Framework adopted at COP-15. PLANGEA is a multicriteria optimization platform that generates a range of scenarios based on goals, targets and objectives such that users can evaluate alternative restoration and land conservation goals related to expanding protected areas. Like WePlan-Forests, PLANGEA uses linear programming to optimize restoration outcomes at national scales. Restoration goals are focused on identifying parcels for conversion to original land use while accounting for biome-specific biodiversity and carbon benefits. The developers state that PLANGEA provides functionality for multi-scale analysis (regional and local) although the authors were unable to determine specifically how this is accomplished. The model itself is not available to the user, but rather the scenarios are pre-simulated by the development staff and made available for exploration on a web interface. Outputs include interactive maps, graphs, tables, images, and reports, and users can download the input data.

PLANGEA integrates the concept of land opportunity cost arising from the loss of agricultural commodity benefits on farmland and cultivated pastures, along with the restoration implementation cost specific to each country. The developers focused on three specific sets of metrics for objectives: 1) Biodiversity conservation measured by the reduction of species extinction debt, the reduction of ecoregions vulnerability, and the reduction of ecosystems integrity impact; 2) Climate change mitigation measured by net values of carbon sequestration (restoration) and emission (conversion); 3) Land opportunity cost for lost benefits from agricultural commodities on farmland and cultivated land.

The PLANGEA platform also serves as a knowledge center for various projects and provides extensive access to publications and related materials. As of this writing the platform continues to be expanded, with work in progress to add metrics permitting multi-criteria optimization for restoration, conservation, and agricultural land conversion.

2.5 Prioritizing Restoration of Sagebrush Ecosystems Tool (PReSET)

The PReSET model [22] was developed specifically to support restoration efforts in the sagebrush (*Artemisia* spp.) steppe ecosystem in western North America [15, 19]. This ecosystem in increasingly impacted by wildfires over several decades and many sagebrush-associated wildlife species are threatened [46]. Restoration activities including seeding and/or planting sagebrush can be economically prohibitive, especially in areas with low ecological resilience [16]. The model authors [22] point out that there is no systematic approach to assessing priority restoration locations within the sagebrush steppe that incorporates (1) multiple restoration approaches and cost), (2) landscape-scale attributes associated with probability of success, and (3) multi-species habitat use and conservation needs.

PReSET has potential application throughout the sagebrush steppe, but as of this writing it has only been evaluated within one key conservation area in southwestern Wyoming, US [22]. PReSET is an R package that links to the PrioritizR package [29] describe above to optimize solutions. The case study data provided by the developers consist of six raster layers that map planning units and attributes for restoration costs, focal species data, and time to recovery after treatment [22]. The planning problem is formulated to select planning units that maximize area with focal species while minimizing the cost of the planning units and time-to-recovery. In the case study, the model application identified specific restoration activities that directly increase sagebrush habitat, as well as the respective costs in the study area. Detailed information on site specific restoration costs and the use of time-to-recovery of sagebrush following oil and gas development sets PReSET apart from many other platforms and applications.

2.6 Forsys

Forsys was developed by USDA Forest Service (hereafter Forest Service) researchers to streamline restoration and wildfire risk reduction prioritization and planning on Western US Forests. The model and case studies are described in a recent report [20] and online [27]. The intended audience for the model includes analysts in federal and state agencies, and the larger research community interested in forest and rangeland planning. There are multiple versions of ForSys including a desktop application written in C++ (ForSysX, described here) and an R [50] version (ForSysR). The Development of ForSys was motivated by the need for spatial planning tools at a range of administrative and operational scales in the Forest Service. Applications to date have ranged from 8,000 ha to over 80 million ha [20] ForSys was primarily designed to support the federal planning process and requirements on US public lands [40] where project alternatives are developed and analyzed using data on costs, revenue, risk, and various measures of resiliency. In a typical application, ForSys is used to identify management activities at the stand polygon scale (1–20 ha) that optimize one or more landscape objectives at the project area scale (5,000–50,000 ha). Management activities include mechanical surface and canopy fuels reduction and burning [32]. ForSysX uses this information to build optimized planning areas using a greedy heuristic where treatments are spatially aggregated into planning areas and optimized for one or more objectives while accounting for spatial constraints and treatment thresholds. Outputs from scenarios have been used in environmental analysis documents to support land management decisions, and in large scale policy documents as well [21, 67].

Common objectives in prior case studies include net revenue from harvesting, resiliency as measured by the stand density index, wildfire risk, and carbon. The program can be used to create a wide range of spatially organized stand treatment patterns, including optimized parcels for broadcast burning, fuel break networks, and landscape restoration. ForSys does not have embedded management response functions, but rather requires that the effects of potential management actions are modeled in advance and included in the input database. For instance, in an economic application, the predicted net revenue is calculated outside of ForSys and presented as choices to the model. The optimization heuristic used in ForSysX is illustrated with online animations [27] and has been contrasted with mathematical programming approaches [1]. The ForSysXR version can be executed from R using a package that calls a ForSys DLL, providing a way to modify landscape conditions within a multiyear planning optimization problem. The ForSysXR package also contains a reporting feature that uses R packages [73] to create interactive reports containing maps, graphics and tabular summaries [5].

2.7 Vibrant Planet Decision Support System (VPDSS)

One of the newer planning systems to emerge to streamline collaborative planning for restoration and wildfire risk reduction in the Western US is VPDSS ([68], Safford et al. in review). VPDSS was created by a team of managers, scientists and technology experts at Vibrant Planet, a public benefits corporation headquartered in California. The development goal is to accelerate collaborative planning [13] by packaging data and planning analytics into a geospatial interface that can optimize management scenarios and quantify project costs and benefits. Although formulating a planning problem is similar to using ForSys in terms of inputs and parameters, VPDSS has a streamlined, cloud hosted interface that facilitates modelling scenarios and visualizing areas of consensus and divergence. VPDSS incorporates cost benefit analyses using risk-based metrics developed from fire simulation models (it can also concurrently incorporate drought risk). Applications to date have ranged from 3,000 to 2 million ha, including selected priority landscapes identified in the Forest Service Wildfire Crisis Strategy [67].

VPDSS, like Forsys, was designed around the recognition that restoration and risk reduction activities are implemented on the ground at the resolution of forest stand polygons (5–50 acres) but need to be organized into project areas at the scale of 5,000–100,000 ha to achieve landscape fire management goals, and to meet logistical and administrative constraints. Optimization, meaning the selection of stands to build a contiguous project area and identification of which stands need to be treated is accomplished with PatchMax algorithm originally developed by the ForSys developers [25]. The heuristic works in a similar fashion to build projects as ForSysX where every stand is evaluated as a potential seed for a patch grown to a specific size. Patches (e.g. stand polygons) are aggregated using Dijkstra's algorithm (Dijkstra 1959) which operates on a distance network of stands to add to patches based on proximity and objective value until activity constraints (e.g., treatment area or cost) are met. As in ForSysX, several parameters control the search behavior to balance the relative importance of distance against objective value when stands are added to a project [25]. The spatial priority weight (SPW) parameter increases the distance between adjacent stands with poor objective quality, thus pushing patches towards areas where the objective values are higher. An SPW setting of 1 penalizes by 10 × the distance cost of adding the stand with the lowest priority score. The exclusion penalty weight (EPW) parameter applies a similar distance penalty if one or both adjacent stands have been excluded from treatment based on the treatment threshold. An EPW setting of 1 penalizes the distance cost by 10 × if it is too costly for Patchmax to search around unavailable stands. The final parameter, the exclusion limit (EL), invalidates patches where the total percent of excluded stands exceeds some predetermined value.

VPDSS is being used with geospatial data derived from a wide range of sources including LiDAR, satellite and aerial imagery, federal agency enterprise geospatial data, and state and local government agencies. The data include metrics that describe forest conditions, values at risk, wildfire and drought risk, management opportunity and other important metrics for scenario planning. Land polygons are developed through a segmentation algorithm done on vegetation structure models based on LiDAR data, with AI methods used to fill LiDAR gaps [68]. Outputs from VPDSS include graphical reports and maps that can be directly incorporated in regulatory planning documents (Safford et al. in review). Tradeoffs among different management scenarios are evaluated by contemporaneously changing the weights for different management objectives (currently eight are available, including fire safety, carbon, biodiversity, asset protection, and recreation) and re-running the optimization. Objectives are normalized in order to reduce bias towards quantity, and normalization may occur relative to local or regional frames of reference. The process provides decisionmakers with information on how specific objectives, including risk reduction to the WUI, affect other goals, like building carbon resilient forests or protecting biodiversity [2, 4]. VPDSS also calculates a rate of return based on costs and benefits using a quantitative wildfire risk framework [26]. Additional outputs include projected treatment costs, economic yields from saw timber and biomass.

2.8 *Planscape*

Planscape is a collaborative development effort among the California Natural Resources Agency, USDA Forest Service, California Wildfire & Forest Resilience Task Force, University of California, Google.org Fellows, and Spatial Informatics Group LLC [47]. Planscape is described as a wildfire resilience decision support tool designed to aid planners in California with design and prioritization of spatially explicit landscape treatments to reduce wildfire risk and maximize ecological benefits under climate change adaptation scenarios. The tool is being designed to facilitate the Shared Stewardship Agreement signed by California Governor Newsom and the US Forest Service to foster cross boundary planning within the State of California. Planscape is funded with public dollars and is available without licensing for public use. The Platform is still under development with prototypes available for use on the website.

The Planscape development approach is similar to VPDSS, where data and analytics are packaged in a cloud hosted environment where a wide range of users can experiment with user-defined planning areas and build optimized treatment plans. The current Planscape interface provides functionality to explore the available data layers and then delineate planning areas and execute simple scenarios based on one management objective at a time. The platform has the potential to facilitate and accelerate collaborative planning efforts that are the backbone of landscape planning in the western US. The scale, resolution, and structure of the planning problem are similar to ForSys where stand polygons, approximated by hexagons of three electable size classes, are organized into project areas and the treated stands optimize one or more objectives subject to activity constraints (e.g. area treated per project). As in VPDSS, optimization is accomplished with the PatchMax algorithm developed by ForSys researchers [25]. Planscape is hosted on a Linux cloud server and development is focused on California landscapes using the Forest Service ACCEL Regional Resource Kit data (California Wildfire and Forest Resilience Task Force 2024). The ACCEL data includes over 200 raster layers that describe forest biophysical conditions in California and a subset of key data themes to use as objectives and constraints. Thus, the scenario analysis functionality of the current version is limited to a predetermined set of priority variables that address wildfire risk and resiliency issues. By integrating data and optimization functionality within a geospatial interface, Planscape offers an open-source solution to develop restoration and risk reduction scenarios for a user community that includes both public and private entities. The development is currently under the direction of the Spatial Informatics Group, a private company, with plans to add a wide range of functionality including fire modelling to evaluate the change in risk from proposed treatments and an array of reporting features.

3 Example Application—VPDSS Planning for Fire Risk Reduction and Ecosystem Restoration

Here we illustrate in more detail an application of the VPDSS planning platform in a real-world case study to establish long-term forest restoration priorities to address wildfire and other management issues on 1.8 million ha of public and private land in the Rogue River Basin, southwest Oregon, USA. The effort is coordinated by the Southern Oregon Forest Restoration Collaborative and funded by the USDA Forest Service as part of a 10-year, $39 million Collaborative Forest Landscape

Restoration Program project. We chose VPDSS for the case study based (1) on its use of multiple technologies and big data to create a system that can streamline complex collaborative planning efforts and generate analyses that can meet the requirements of federal regulatory planning, and (2) its implementation on similar collaborative large-landscape planning efforts across the western US, including many of the Forest Service Wildfire Crisis Landscapes [67].

The broad land management goals for this project characterize typical socio-ecological complexity for a large multi-owner tract of land in the western US, and include: (1) maintain and restore diverse habitats, species, and ecosystem structures; (2) increase forest resistance and resilience to fire, insects, and drought; (3) conserve old trees and stands within and outside complex forest habitat; (4) establish conditions for under-burning to safely maintain landscape resilience and fire suppression capabilities; and (5) generate ongoing products and employment through long-term forest maintenance and timber harvest consistent with principals of ecological forestry [38]. The problem at hand is to prioritize and schedule active forest and fuel management prescriptions [32] that will effect changes in landscape conditions to meet these goals in varying balances, as prescribed by the users.

SOFRC used Marxan for the first round of project planning, and project partners were generally satisfied with the results. However, Marxan expertise was restricted to a single member of the supporting staff, and project members expressed a desire for a more user-friendly platform that supported collaborative scenario building and could be used by multiple stakeholders independent of their computer skills, and a data update cycle that permitted rapid adaptation and re-prioritization (K. Metlen, The Nature Conservancy, pers. comm.). SOFRC contracted with Vibrant Planet in 2022 to develop an actionable restoration plan including data updates as needed to account for changing landscape conditions from management and wildfire. Various rounds of scenario development were carried out with project stakeholders, for various subsets of the SOFRC area.

As an example of this process, we built management scenarios for a 35,000-ha subregion of the SOFRC area that included the city of Ashland and which also supports the highest density of high risk properties in the state of Oregon [44]. Three divergent scenarios were created based on typical stakeholder objectives in the SOFRC project. In VPDSS, users interact and provide input to scenarios on the geospatial interface using slider bars to change weights for each of eight management objective categories (Fig. 1). The first "Environmental" scenario used a weighted objective function with relatively high weights for management actions that would protect biodiversity, carbon, and water, and to a lesser extent recreation facilities (Fig. 1). The second, "Fire Safety and Commodity" scenario gave highest weighting to Asset Protection, Fire Safety, and Commodities (Fig. 1). A third "Equal Weighting" scenario equally weighted all of the objectives (see Safford et al., in review). Costs and revenue associated with each potential management action (different categories of thinning, burning, etc.) were compiled and assigned to each treatment parcel in the study area. The scenario analyzed in VPDSS optimized RROI for a five project plan that treated a maximum of 1000 ha per project with a maximum budget per project of $1,000,000 (Fig. 1). Optimization required 8–10 s per scenario and the results were rendered into consensus categories to facilitate collaborative discussions (Fig. 1). The consensus categories are based on the treatment overlap between the scenarios: complete, majority, or none. RROI for each objective and scenario are displayed in the outputs allowing users to see the attainment towards the various objectives for a given scenario, along with the consensus scenario with the treated area common to all three (complete consensus) or two of the three scenarios (majority consensus). The results showed that the scenarios diverged spatially (selected non-overlapping treatment parcels) in addressing different stakeholder interests (Fig. 1 A,B) and that achieving consensus required shifting treatments from federal to local government ownerships. The complete consensus lands also had the highest treatment costs per ha. The application of VPDSS in a multi-objective planning environment resulted in several observations from project stakeholders and developers. For example, the efficiency of the VPDSS platform and support by Vibrant Planet did much to "automate the standard, plodding resource management planning process" and sped things up considerably. In addition, the more transparent analytical process and client-friendly interface was key to efficiently managing a large collaborative group with many engaged but non-technical partners (T. Fairbanks, SOFRC, perc. comm.).

4 Discussion

The goal of this paper was to examine newer developments in terrestrial spatial planning platforms and provide a perspective on their application and functionality to solve a wide range of spatial planning problems in natural resource conservation and management. Similar decision support systems are also widely used in marine spatial planning and the reader is referred elsewhere for similar comparisons [33]. The planning platforms discussed here integrate a range of newer technologies including cloud hosting, big data, geospatial interfaces, and spatial optimization to identify cost effective solutions to address wide ranging conservation, restoration, and natural hazard problems. However, despite this broad similarity, the reviewed platforms differ in many significant ways, especially in the wide range of configurations, ranging from desktop executables with interfaces for specifying inputs and outputs, to cloud-based geospatial interfaces that provide for exploration of spatial data and formulation/execution of scenarios. Beyond the technology aspect, when conservation platforms like Marxan, Plangea and WePlan-Forests were compared to ForSys, VPDSS, and Planscape it was apparent that different spatial planning problems are being solved. Many of the conservation platforms optimize reserve design by one or more passive management actions including parcel acquisition, protective regulations, or incentives to return lands to uses that are compatible with conservation goals. The optimization problem is formulated to maximize conservation target values assigned to decision units (generally referred to planning areas in the literature) with penalties related to boundary lengths or other parcel scale features to enforce some level of connectivity. A few exceptions to this generalization include applications where PReSet and Marxan were used to optimize active management,

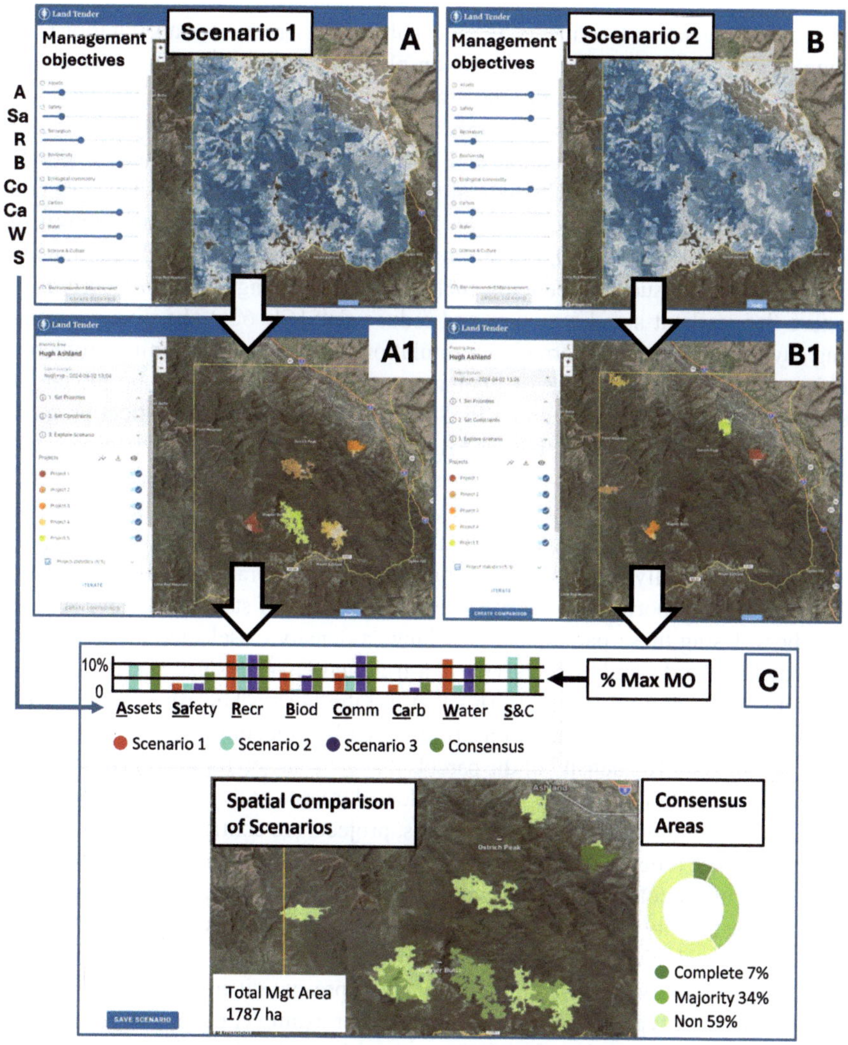

Figure. 1. Example outputs from three VPDSS scenarios for a 35,000-ha landscape in the south-eastern SOFRC project area: an "Environmental" scenario (Scenario 1), a "Fire Safety and Commodity" scenario (Scenario 2) scenario, and an "Equal Weighting" scenario (not shown). Each scenario consisted of five 1000-ha projects optimized for investment returns based on a mixed objective function, with the weightings given to each Management Objective category and the potential RROI mapped in Panels A and B (see text). Panels A1 and B1 show the location of the five optimized projects for Scenarios 1 and 2. Panel C shows the returns for each objective RROI by scenario (% of maximum potential RROI attained for each Management Objective category), and also a map of spatial project consensus among the three scenarios. "Complete" = all three scenarios agreed; "majority" = 2 of 3 scenarios agreed; "non" = scenarios did not agree. Under Management Objectives: A = Asset protection, Sa = Fire safety, R = Recreation, B = Biodiversity, Co = Commodities, Ca = Carbon, W = Water, S = Science and culture. Additional details in Safford et al. (in review)

i.e. specific vegetation treatments, to meet habitat conservation goals (citation). The planning process using these conservation platforms are carried out under diverse set of actors, government institutions, stakeholders and ecosystems.

By contrast, platforms like VPDSS and ForSys were specifically designed to support analyses in risk reduction and restoration on US public lands under the umbrella of a well-defined US federal regulatory planning established under the National Environmental Policy Act [66]. Here, models are used to build optimized project areas (e.g. 10,000–40,000 ha) to prioritize specific forest and fuel management investments, such as reducing fuel loadings (thinning, broadcast burning) to restore fire resilient stand structure and reduce risk [61, 62]. The divergent planning environments led to alternative formulations of the spatial planning problem to achieve connectivity of stands or planning areas between the heuristic models we reviewed (ForSys, Planscape and VPDSS) and the various mathematical programming (MP) models. Platforms that use MP solvers all use a "top-down" approach with global constraints or penalties to achieve some level of connectivity and compactness of the solution. For instance, PrioritizR uses two parameters to encourage clumped solutions, one being a boundary penalty that scales the importance of selecting polygons that are spatially clumped versus maximizing the overall objective, the other is an edge factor proportion that encourages the selection of stands that have selected neighbors. Using these parameters to achieve a density of selected planning areas that matches the desired connectivity is a trial-and-error process [1]. By contrast, connectivity in "bottom-up" platforms that use the ForSys heuristic is enforced by an algorithm that adds parcels to a project area by selecting candidates using an adjacency matrix. The density of the parcels that are selected for management within the project area is controlled by the user-selected treatment threshold parameter that describes the conditions in the parcels. Thus, projects are contiguous stands, and the final project size is determined by the interplay of a treatment threshold (if specified), the density of parcels that exceed that threshold, the objective contribution, and a project-scale activity constraint on total area treated or cost. This bottom-up approach to recursively colonize landscapes with projects versus the top-down approach that controls connectivity with boundary penalties represent divergent spatial formulations for different types of planning problems. For example, generating a multiyear action plan with a sequence of projects, each with a defined boundary, is most efficiently accomplished through a recursive bottom-up optimization, where successive projects are optimized considering only the remaining landscape. Replicating the ForSys or VPDSS spatial optimization planning framework with top-down platforms like Marxan, WePlan, PrioritizR, and PreSet would require defining planning areas or zones beforehand, in which case exact optimal solutions could be found more easily simply by sorting the decision units based on their objective value [1]. Other than VPDSS [68] and Planscape [47], a planning system that currently replicates the functionality of ForSys is not apparent from the literature, despite the recent appearance of several new platforms discussed here [22, 31, 63]. Another key difference between the platforms is that models like ForSys, Planscape, and VPDSS were developed with an emphasis for wide application by field planners and for collaborative planning where transparency is key, and they use a relatively simple greedy spatial heuristic instead of more complicated mathematical programming approaches that are gaining traction in the literature [57]. Recent comparisons of mathematical programming with heuristic solutions and found the former to consistently provide a superior approach in terms of achieving exact optima [57] and MP solvers are now being used with Marxan and always have been with PrioritizR and PReSET. When paired with an efficient optimizer [28], there is evidence that extremely large spatial problems can now be solved much more rapidly than previously possible [57]. While mathematical programming methods are scientifically more elegant and can achieve exact optima [9], their application to the problem of restoration and risk reduction prioritization on US public lands has mostly been in the research domain and to our knowledge there is no recent field application to support actual management decision-making [1, 20] On the other hand, ForSys has had widespread use in forest management applications [27] and the ForSys-based VPDSS platform is quickly becoming a fundamental tool for restoration and risk reduction prioritization on multiple large landscapes in the western US (Safford et al. in review). Our review found it difficult to document actual land management decisions resulting from the application of many of the other spatial planning and optimization models described as decision support systems [8].

5 Conclusions

Prioritization platforms are now available for a wide range of ecosystems and management problems, and leverage many new technologies to address terrestrial restoration, risk reduction, and biodiversity conservation. In the case of restoring fire-adapted forests in the western US managed by public land management agencies, spatially explicit scenario analyses, whether at the watershed or national scale, are the blueprints to communicate conservation and restoration plans to government budget appropriations committees and local stakeholder groups that engage in project planning [23, 54]. Plans that prioritize investments in active forest management communicate specific budget and workforce capacity needs to achieve expected outcomes in space and time. Careful prioritization of investments in forest restoration and risk reduction activities is necessary to capture and properly respond to variation in landscape conditions and risk factors relative to management goals [12].

The number of decision support platforms for ecosystem restoration, risk reduction, and conservation is growing rapidly, as is their sophistication. Today, most of the advances are happening outside of government research institutions and they are increasingly being supported by private funding. With the exception of the military, the public sector has generally been slow in adopting new technology, mostly due to lack of funding, low levels of information-technology, concerns about data integrity, and organizational decentralization [30]. Multiple reports and assessments have concluded that (1) the speed and scale of global change requires the rapid adoption of innovative digital data and tools; (2) expanded and intensified public-private partnerships are fundamental to bridge the science-management gap; and (3) much of the science and technology needed to support restoration and risk reduction efforts already exist today but aren't being used by the public sector [30]. At this critical time, uptake of decision support platforms and underlying science by public and private entities is not keeping pace with the rate of technology development. We conclude that bottlenecks in application and realizing potential benefits rest more with institutional capacity to absorb and implement these technologies than the design, development, and deployment of them.

Acknowledgements We thank anonymous reviews for helpful comments on an earlier draft of the manuscript. Bruno Aparicio, Michelle Day, and Cody Evers contributed to useful discussions on improving the design of decision support systems to meet the needs of operational planners. Partial funding for this research was provided by Vibrant Planet.

Literature cited

1. Ager, A.: Improving the evaluation of spatial optimization models for prioritizing landscape restoration and wildfire risk reduction investments. J. Environ. Manage. **360**, 121001 (2024)
2. Ager, A.A., Day, M.A., Vogler, K.: Production possibility frontiers and socioecological tradeoffs for restoration of fire adapted forests. Journal of Environmental Management 176:157–168. https://www.fs.usda.gov/research/treesearch/52994. (2016)
3. Ager, A.A., Evers, C.R., Day, M.A., Alcasena, F.J., Houtman, R.: Planning for future fire: scenario analysis of an accelerated fuel reduction plan for the western United States. Landsc. Urban Plan. **215**: 104212. https://doi.org/10.1016/j.landurbplan.2021.104212 https://www.fs.usda.gov/treesearch/pubs/63129. (2021)
4. Ager, A.A., Houtman, R., Day, M.A., Ringo, C., Palaiologou, P.: Tradeoffs between US national forest harvest targets and fuel management to reduce wildfire transmission to the wildland urban interface. For. Ecol. Manag. **434**, 99–109. https://www.fs.usda.gov/treesearch/pubs/57897. (2019)
5. Aparício, B.A., Bunzel, K., Day, M., Ager, A.: ForSysXR: a R package to execute C++ (ForSysX) version of the ForSys scenario planning model. R package version 0.9. Available on request from github. (2024)
6. Ball, I.R. Possingham, H.P.: MARXAN. v1.8.2: marine reserve design using spatially explicit annealing (2000)
7. Ball, I.R., Possingham, H.P., Watts, M.: Marxan and relatives: software for spatial conservation prioritisation. In: Moilanen, A., Wilson, K.A., Possingham, H.P. (eds.) Spatial Conservation Prioritisation: Quantitative Methods and Computational Tools, pp. 185–210. Oxford University Press, Oxford (2009)
8. Baskent, E.Z., Borges, J.G., Kašpar, J.: An updated review of spatial forest planning: approaches, techniques, challenges, and future directions. Curr. For. Rep. 1–23 (2024)
9. Beyer, H.L., Dujardin, Y., Watts, M.E., Possingham, H.P.: Solving conservation planning problems with integer linear programming. Ecol. Model. **328**, 14–22 (2016)
10. Biggs, R., Raudsepp-Hearne, C., Atkinson-Palombo, C., Bohensky, E., Boyd, E., Cundill, G. et al.: Linking futures across scales: a dialog on multiscale scenarios. Ecol. Soc. 12 (2007)
11. Brudvig, L.A.: Toward prediction in the restoration of biodiversity. J. Appl. Ecol. **54**, 1013–1017 (2017)
12. Brudvig, L.A., Barak, R.S., Bauer, J.T., Caughlin, T.T., Laughlin, D.C., Larios, L., Matthews, J.W., Stuble, K.L., Turley, N.E., Zirbel, C.R.: Interpreting variation to advance predictive restoration science. J. Appl. Ecol. **54**, 1018–1027 (2017)

13. Butler, W.H., Esch, B.: Collaborative forest landscape restoration in action. In: Butler, W.H., Schultz, C.A. (eds.) A New Era for Collaborative Forest Management: Policy and Practice Insights From the Collaborative Forest Landscape Restoration Program. Routledge, London (2019)
14. Cao, K., Li, W., Church, R.: Big Data, Spatial Optimization, and Planning, pp. 941–947. SAGE Publications Sage UK, London, England (2020)
15. Chambers, J.C., Brown, J.L., Bradford, J.B., Board, D.I., Campbell, S.B., Clause, K.J., Hanberry, B., Schlaepfer, D.R., Urza, A.K.: New indicators of ecological resilience and invasion resistance to support prioritization and management in the sagebrush biome, United States. Front. Ecol. Evol. **10**, 1009268 (2023)
16. Chambers, J.C., Maestas, J.D., Pyke, D.A., Boyd, C.S., Pellant, M., Wuenschel, A.: Using resilience and resistance concepts to manage persistent threats to sagebrush ecosystems and greater sage-grouse. Rangel. Ecol. Manage. **70**, 149–164 (2017)
17. Chazdon, R.L., Guariguata, M.R.: Decision support tools for forest landscape restoration: current status and future outlook. Occasional paper 183, center for international forestry research, Indonesia (2018)
18. Cleland, D., Reynolds, K., Vaughan, R., Schrader, B., Li, H., Laing, L.: Terrestrial condition assessment for national forests of the USDA forest service in the continental US. Sustainability **9**, 2144 (2017). https://doi.org/10.3390/su9112144
19. Crist, M.R., Chambers, J.C., Phillips, S.L., Prentice, K.L., Wiechman, L.A. (eds.): Science framework for conservation and restoration of the sagebrush biome: linking the department of the interior's integrated rangeland fire management strategy to long-rerm strategic conservation actions. Part 2. Management applications. Gen. Tech. Rep. RMRS-GTR-389. USDA, forest service, rocky mountain research station, fort collins, CO (2019)
20. Day, M.A., Ager, A.A., Houtman, R., Evers, C.R.: Prioritizing restoration and risk reduction landscape projects with the ForSys planning system. Gen. Tech. Rep. RMRS-GTR-437, USDA Forest service, rocky mountain research station, Fort Collins, CO. https://doi.org/10.2737/RMRS-GTR-437. (2023)
21. Day, M.A., Houtman, R., Belavenutti, P., Ringo, C., Ager, A.A., Bassett, S.: An assessment of forest and woodland restoration priorities to address wildfire risk in New Mexico. Gen. Tech. Rep. RMRS-GTR-423, USDA forest service, rocky mountain research station, Fort Collins, CO. https://doi.org/10.2737/RMRS-GTR-423. (2021)
22. Duchardt, C.J., Monroe, A.P., Heinrichs, J.A., O'Donnell, M.S., Edmunds, D.R., Aldridge, C.L.: Prioritizing restoration areas to conserve multiple sagebrush-associated wildlife species. Biol. Cons. **260**, 109212 (2021)
23. Eaton, M.J., Yurek, S., Haider, Z., Martin, J., Johnson, F.A., Udell, B.J., Charkhgard, H., Kwon, C.: Spatial conservation planning under uncertainty: adapting to climate change risks using modern portfolio theory. Ecol. Appl. **29**, e01962 (2019). https://doi.org/10.1002/eap.1962
24. Esfandeh, S., Kaboli, M., Eslami-Andargoli, L.: A chronological review on application of MARXAN tool for systematic conservation planning in landscape. International Journal of Engineering and Applied Sciences **2**, 6–17 (2015)
25. Evers, C., Belavenutti, P., Day, M.A., Ager, A.A., Houtman, R.: patchmax: a spatial optimization package for building spatial priorities. Version 0.2.0. Available at: https://github.com/forsys-sp/patchmax. (2023)
26. Finney, M.A.: The challenge of quantitative risk analysis for wildland fire. For. Ecol. Manag. **211**, 97–108. <Go to ISI>://WOS:000229875600010. (2005)
27. ForSys Research Consortium.: The forsys research consortium. www.forsysplanning.org. (2024)
28. Gurobi Optimization, L.: Gurobi optimizer reference manual. Version. Available at: https://www.gurobi.com. (2023)
29. Hanson, J.O., Schuster, R., Morrell, N., Strimas-Mackey, M. Edwards, B.P.M., Watts, M.E., Arcese, P., Bennett, J.R., Possingham, H.P.: Prioritizr: systematic conservation prioritization in R. R package version 8.0.4.1. https://github.com/prioritizr/prioritizr, https://prioritizr.net. (2024)

30. Hinkley, S:. Technology in the Public Sector and the Future of Government Work. University of California Labor Center, Berkeley, CA. https://laborcenter.berkeley.edu/technology-in-the-public-sector-and-the-future-of-government-work/. (2023)
31. IIS.: WePlan forests: forest restoration planning. http://weplan-forests.org/index.html. (2024)
32. Jain, T.B., Battaglia, M.A., Han, H.-S., Graham, R.T., Keyes, C.R., Fried, J.S., Sandquist, J.E.: A comprehensive guide to fuel management practices for dry mixed conifer forests in the northwestern United States. Gen. Tech. Rep. RMRS-GTR-292, USDA forest service, rocky mountain research station, fort collins, CO. http://www.fs.fed.us/rm/pubs/rmrs_gtr292.pdf. (2012)
33. Janßen, H., Göke, C., Luttmann, A.: Knowledge integration in marine spatial planning: a practitioners' view on decision support tools with special focus on Marxan. Ocean Coast. Manag. **168**, 130–138 (2019)
34. Keeley, J.E., Safford, H., Fotheringham, C.J., Franklin, J., Moritz, M.: The 2007 southern California wildfires: lessons in complexity. J. Forest. **107**, 287–296 (2009)
35. Kharazi Baheri, B.: MARXAN conservation planning decision support software. Zist Sepehr Stud. Mag. **10**, 58–61 (2015)
36. Kukkala, A.S., Moilanen, A.: Core concepts of spatial prioritisation in systematic conservation planning. Biol. Rev. **88**, 443–464 (2013)
37. Matsuba, M., Akasaka, M., Miyashita, T.: Marxan for systematic conservation planning: principles and applications. Jpn. J. Conserv. Ecol. **20**, 35–47 (2015)
38. Metlen, K.L., Fairbanks, T., Bennett, M., Volpe, J., Kuhn, B., Thompson, M.P., Thrailkill, J., Schindel, M., Helmbrecht, D., Scott, J.: Integrating forest restoration, adaptation, and proactive fire management: rogue river basin case study. Can. J. For. Res. **51**, 1292–1306 (2021)
39. Moilanen, A., Leathwick, J.R., Quinn, J.M.: Spatial prioritization of conservation management. Conserv. Lett. **4**, 383–393 (2011)
40. NEPA.: National environmental policy act. 42 U.S.C., USA. (1969)
41. Nofal, O.M., Van De Lindt, J.W., Do, T.Q., Yan, G., Hamideh, S., Cox, D.T., Dietrich, J.C.: Methodology for regional multihazard hurricane damage and risk assessment. J. Struct. Eng. **147**, 04021185 (2021)
42. Noonan-Wright, E., Seielstad, C.A.: Patterns of wildfire risk in the United States from systematic operational risk assessments: how risk is characterised by land managers. Int. J. Wildland Fire **30**, 569 (2021)
43. Noss, R., Nielsen, S., Vance-Boland, K.: Chapter 12—prioritizing ecosystems, species and sites for restoration. In: Moilanen, A., Wilson, K.A., Possingham, H.P. (eds.) Spatial Conservation Prioritization: Quantitative Methods and Computational Tools, pp. 158–170. Oxford University Press, Oxford (2009)
44. Oregon Wildfire Risk Explorer. https://oregon-explorer.apps.geocortex.com/webviewer/?app=665fe61be984472da6906d7ebc9a190d&viewer=wildfire)
45. Peterson, G.D., Cumming, G.S., Carpenter, S.R.: Scenario planning: a tool for conservation in an uncertain world. Conserv. Biol. **17**, 358–366 (2003)
46. Pilliod, D.S., Jeffries, M.A., Welty, J.L., Arkle, R.S.: Protecting restoration investments from the cheatgrass-fire cycle in sagebrush steppe. Conserv. Sci. Pract. **3**(10), 508 (2021)
47. Planscape.: Planscape: a planning tool to maximize wildfire resilience and ecological benefits. https://www.planscape.org/. (2024)
48. Pohjanmies, T., Eyvindson, K., Mönkkönen, M.: Forest management optimization across spatial scales to reconcile economic and conservation objectives. PLoS ONE **14**, e0218213 (2019)
49. Pressey, R.L., Bottrill, M.C.: Approaches to landscape-and seascape-scale conservation planning: convergence, contrasts and challenges. Oryx **43**, 464–475 (2009)
50. R Core Team.: R: a language and environment for statistical computing. Version 3.5.3. Available at: http://www.R-project.org/. (2019)
51. Reinhardt, J.R., Tack, J.D., Maestas, J.D., Naugle, D.E., Falkowski, M.J., Doherty, K.E.: Optimizing targeting of pinyon-juniper management for sagebrush birds of conservation concern while avoiding imperiled pinyon jay. Rangel. Ecol. Manage. **88**, 62–69 (2023)

52. Reynolds, K.M., Hessburg, P.F., Keane, R.E., Menakis, J.P.: National fuel-treatment budgeting in US federal agencies: capturing opportunities for transparent decision-making. For. Ecol. Manage. **258**, 2373–2381 (2009)
53. Reynolds, K.M., Twery, M., Lexer, M.J., Vacik, H., Ray, D., Shao, G., Borges, J.G.: Decision support systems in forest management. In: Handbook on Decision Support Systems 2, pp. 499–533. Springer (2008)
54. Riddell, G.A., van Delden, H., Maier, H.R., Zecchin, A.C.: Exploratory scenario analysis for disaster risk reduction: considering alternative pathways in disaster risk assessment. Int. J. Disaster Risk Reduct. **39**, 101230 (2019). https://doi.org/10.1016/j.ijdrr.2019.101230doi:10.1016/j.ijdrr.2019.101230
55. Safford, H.D., Miller, C., Perrot, D., Duffy, K., Gilbert, S., Hoecker, T., Koontz, M., Thompson, M., Shannon, J., Scott, J., Conway, S.: In review. a collaborative, cloud-based decision support system for wildfire risk mitigation and ecosystem restoration. Ecological Modelling.
56. Sarkar, S., Pressey, R.L., Faith, D.P., Margules, C.R., Fuller, T., Stoms, D.M., Moffett, A., Wilson, K.A., Williams, K.J., Williams, P.H.: Biodiversity conservation planning tools: present status and challenges for the future. Annu. Rev. Environ. Resour. **31**, 123–159 (2006)
57. Schuster, R., Hanson, J.O., Strimas-Mackey, M., Bennett, J.R.: Exact integer linear programming solvers outperform simulated annealing for solving conservation planning problems. PeerJ **8**, e9258 (2020). https://doi.org/10.7717/peerj.9258doi:10.7717/peerj.9258
58. SPARC.: Spatial planning for area conservation in response to climate change https://www.sparc-website.org/). (2019)
60. Star, J., Rowland, E.L., Black, M.E., Enquist, C.A.F., Garfin, G., Hoffman, C.H., Hartmann, H., Jacobs, K.L., Moss, R.H., Waple, A.M.: Supporting adaptation decisions through scenario planning: enabling the effective use of multiple methods. Clim. Risk Manag. **13**, 88–94 (2016)
61. Stephens, S.L., Collins, B.M., Biber, E., Fulé, P.Z.: US federal fire and forest policy: emphasizing resilience in dry forests. Ecosphere **7**, e01584 (2016). https://doi.org/10.1002/ecs2.1584doi:10.1002/ecs2.1584
62. Stephens, S.L., Foster, D.E., Battles, J.J., Bernal, A.A., Collins, B.M., Hedges, R., Moghaddas, J.J., Roughton, A.T., York, R.A.: Forest restoration and fuels reduction work: different pathways for achieving success in the S ierra N evada. Ecol. Appl. e2932 (2023)
63. Strassburg, B.B., Iribarrem, A., Beyer, H.L., Cordeiro, C.L., Crouzeilles, R., Jakovac, C.C., Braga Junqueira, A., Lacerda, E., Latawiec, A.E., Balmford, A.: Global priority areas for ecosystem restoration. Nature **586**, 724–729 (2020)
64. Trammell, E.J., Thomas, J.S., Mouat, D., Korbulic, Q., Bassett, S.: Developing alternative land-use scenarios to facilitate natural resource management across jurisdictional boundaries. J. Environ. Planning Manage. **61**, 64–85 (2018)
65. USDA Forest Service.: Watershed condition framework: a framework for assessing and tracking changes to watershed condition. FS-977. http://www.fs.fed.us/publications/watershed/. (2011)
66. USDA Forest Service.: National Environmental Policy Act (NEPA). https://www.fs.fed.us/emc/nepa/. (2020)
67. USDA Forest Service.: Confronting the wildfire crisis: a strategy for protecting communities and improving resilience in america's forests. FS-1187a, Washington D.C. https://www.fs.usda.gov/sites/default/files/fs_media/fs_document/Confronting-the-Wildfire-Crisis.pdf. (2022)
68. Vibrant Planet.: Land tender: a platform for landscape restoration. https://www.vibrantplanet.net/landtender. (2023)
69. Villarreal-Rosas, J., Sonter, L.J., Runting, R.K., López-Cubillos, S., Dade, M.C., Possingham, H.P., Rhodes, J.R.: Advancing systematic conservation planning for ecosystem services. Trends Ecol. Evol. **35**, 1129–1139 (2020)
70. Wade, C.M., Austin, K.G., Cajka, J., Lapidus, D., Everett, K.H., Galperin, D., Maynard, R., Sobel, A.: What is threatening forests in protected areas? a global assessment of deforestation in protected areas, 2001–2018. Forests **11**, 539. https://www.ncbi.nlm.nih.gov/pubmed/33123319. (2020)
71. Watts, M.E., Stewart, R.R., Martin, T.G., Klein, C.J., Carwardine, J., Possingham, H.P.: Systematic conservation planning with Marxan. In: Learning Landscape Ecology: A Practical Guide to Concepts and Techniques, pp. 211–227. (2017)

72. Williams, B.A., López-Cubillos, S., Ochoa-Quintero, J.M., Crouzeilles, R., Villa-Piñeros, M., Isaacs Cubides, P.J., Schmoeller, M., Marin, W., Tedesco, A., Bastos, D.: Bringing the forest back: restoration priorities in Colombia. Divers. Distrib. **30**, e13821 (2024)
73. Xie, Y., Allaire, J.J., Grolemund, G.: R markdown: the definitive guide. Chapman and Hall/CRC. (2018)
74. Zhang, L., Li, J.: Identifying priority areas for biodiversity conservation based on Marxan and InVEST model. Landscape Ecol. **37**, 3043–3058 (2022)

pyO3F—A Python Framework for Wildfire-Related Optimization, Part I: Design and Fundamentals

Filipe Alvelos

Abstract The purpose of the framework pyO3F is to support the development of optimization approaches, such as meta-heuristics or mixed integer programming, for wildfire-related optimization problems. Given an extent, it allows the definition of scenarios that a user may use as a base to his approaches and/or directly to estimate fire potential and/or simulation of fire spread. The main contribution is the consideration of fire-related resources and their interaction with fire. pyO3F is based on the minimum travel time principle that states that fire follows the quickest path between any two locations in a landscape. This principle makes it possible to model fire spread as a set of shortest paths in a network. The actual or potential positions and movements of the firefighting resources are defined in networks that interact with the fire network according to their type (for example, suppression or detection). This is the first paper of two on pyO3F. In this paper, we describe the architecture of pyO3F and how fire, resources, and their interaction are modeled and implemented. In the second paper, pyO3F—A python framework for wildfire-related optimization, Part II: Usage, we describe how it can be used to provide fire scenarios, to estimate fire spread and fire potential, with or without the presence of resources. pyO3F is one of the results of the project *O3F - An Optimization Framework to Reduce Forest Fires*.

Keywords Software · Ffire spread simulation · Wildfire supression · Mixed integer programming

1 Introduction

In this paper, we describe a computational framework, pyO3F, that supports the combination of fire spread and optimization. pyO3F is a set of python modules / classes that characterizes the landscape and the available resources and allows generating scenarios (based on fuels, wind, and slope) that may include an initial

F. Alvelos (✉)
Department of Production and Systems/ALGORITMI Research Center/LASI, School of Engineering, University of Minho, Braga, Portugal
e-mail: falvelos@dps.uminho.pt

F. Alvelos et al. (eds.), *Advances in Optimization and Wildfire*, Lecture Notes in Operations Research, https://doi.org/10.1007/978-3-032-03108-2_3

state, characterized by one or more ignitions or by one or more fire perimeters. Scenarios can also be generated through random parameters (e.g., wind direction following a given distribution).

Through simple functions, it is possible to simulate fire spreads, estimate fire potentials, and use pyO3F scenarios in optimization approaches. Fire potential measurements are obtained by aggregating the results of a set of fire spread simulations for independent ignitions. An example of a fire potential measure is the number of fire paths with a Rate Of Spread (ROS) above a given threshold. These topics are detailed in the second paper devoted to pyO3F [1] in this same volume.

In this paper we describe the design and fundamentals of pyO3F: how the extent of interest is represented, how fire behaviour is modelled, and how different types of resources are characterized and their relation with fire.

pyO3F uses extensively the package *networkx* [2] and also *geopandas* [3] and *pandas* [4], *numpy* [5], *matplotlib* [6], *shapely* [7], and *rasterio* [8]. For solving mixed integer programming (MIP) models, it uses Gurobi through *gurobipy* [9]. pyO3F can be used for any (rectangular) extent in Portuguese mainland, for which public data on land use and altitudes are available. Fuels models can also be inputted. The output is provided in text files and shape files that can be analyzed and further processed in a Geographic Information System such as QGIS [10].

Although a large number of software for fire spread simulation, or related topics, exist (see next section), to our best knowledge, none is devoted to support fire-related optimization in the strict sense (e.g., meta-heuristics and MIP approaches). pyO3F aims to contribute to reducing this gap between optimization and wildfire management.

In Sect. 2, we frame pyO3F in the main categories of existing fire-related approaches and refer to some of the most relevant software for each. In Sect. 3, we provide an overview of the pyO3F architecture. The core classes are detailed in Sect. 4. Lastly, in Sect. 5, we summarize the main conclusions of this work.

2 Related Work

There are multiple fire-related software that implements fire spread models. Sullivan [11–13] provides a comprehensive survey of fire spread models and their implementations, classified into physical, quasi-physical, empirical, quasi-empirical, mathematical analogues, and simulation models. More than 40 models are described.

Particularly appropriate for fire spread at a landscape level, the scale which pyO3F addresses, are simulation models. In [14] a description of 23 fire spread simulators is conducted and *Farsite* [15] is described in depth. More recently, Ghodrat et al. [16] survey 14 simulators. Cardil et al. [17] provides an overview of the main challenges of software for decision-making based on fire spread simulation.

Approaches implemented in the most used software for wildfire spread simulation can be divided in three major groups: based on wave propagation (Huygens' principle), cellular automata and network-based. The former two are based on time discretization, while in the latter, time is continuous. The latter two are based on space discretization (also referred to as raster-based), while the former is based on defining the fire perimeter as a curve.

In models based on wave propagation, at each time step, an ellipse is defined for each point of a set of points defining the perimeter. The size of the ellipse depends on the fuel, and the shape and orientation depend on the wind-slope. The (updated) perimeter at the end of the timestep is the envelope of all ellipses. In [15], a detailed description of the use of the Huygens principle in wildfire spread is given. In [14], it is mentioned that two software systems stands out of the 23 tested: *FARSITE* [15] and *Prometheus* [18], both based on wave propagation. The former was developed by the US Forest Service, and the latter is the Canadian wildland fire growth simulation model.

Cellular automata are also based on a time discretization. In contrast to wave propagation methods, a spatial discretization is conducted *a priori* resulting in a set of cells (usually a grid). At each timestep, each cell has a state from a set of finite states. Typical states are unburnable, unburned, burning, and burned. Each time step may change the status of some cells according to transition rules that take into account the state of neighbor cells. Two examples of cellular automata approaches are [19] and [20]. In the former, the spread of the fire is modeled with detailed geographical and meteorological data. The probability of a cell being ignited by an adjacent cell is modeled by the density, type, and height of vegetation, and also by the wind, slope of the terrain, and moisture content. In [20], a machine learning component is included in the transition rules.

A software based on cellular automata is *cell2fire* [21]. In [22], this simulator is extended to accommodate additional fuel models and crown fires. Also *Propagator* [23] is based on cellular automata. A major difference between these two approaches is that the former uses deterministic transition rules based on the generally accepted elliptical shape of the fire, and the latter uses transition rules based on fire transmission probabilities.

Network models represent the landscape as a graph: nodes are associated with locations and arcs with potential fire transmission between adjacent locations. Fire spread is modeled according to the Minimum Travel Time (MTT) principle which states that the fire follows the fastest path between two locations in the landscape [24]. Associating fire transmission times with arcs, it is possible to estimate fire spread at the landscape level through shortest path algorithms, which are particularly efficient [25]. For an overview of network approaches, we refer the interested reader to [26].

Flammap [27], that now includes *Farsite*, developed by the US Forest Service, one of the most used software, is based on MTT. Also, Wildfire Analyst [28] from TechnoSylva is based on MTT. pyO3F is based on the MTT principle, which can be implemented through well-known optimization algorithms (e.g. Dijkstra's algorithm) or linear programming. Using MTT, a common ground emerges that supports both fire modeling and optimization. For example, a single Mixed Integer Programming (MIP) model may include decision variables for fire arrival times and decision variables for the positioning of suppression resources.

Some of these software offer features related to the presence of resources and / or decision making. For example, in *Propagator* fire fighting actions are taken into account from historic records but not optimized. Another example is the heuristic of *Flammap* for deciding fuel treatments. A main characteristic of pyO3F is that it was developed from the start to support optimization approaches (mathematical programming and metaheuristics) with a strong emphasis on modelling firefighting resources. Those are positioned and/or moved in their own network, with their own attributes (e.g. speed), which is distinct from the fire network (i.e. the MTT network).

3 Overview

The pyO3F Python framework consists of a set of classes that, for a given landscape, allow (i) to simulate fire spread with one or more ignitions, (ii) to estimate the fire potential of a landscape by considering multiple independent ignitions, and (iii) to implement optimization approaches for fire-related problems based on the previous two points or based on scenarios provided by pyO3F. In these three uses, approaches based on the pyO3F, may include firefighting resources, their potential locations and movements over time, and their interactions with fire. In this section, we provide an overview of pyO3F. In the following, we will detail the main components.

Figure 1 provides an overview of pyO3F. The blocks correspond to data (extent, resolution, fuel models) or classes (all the others). We differentiate blocks by whether they are related to inputs (light gray, optional if with dashed border), outputs (dark gray), or are internal (in white). The block `scenario`, in black, has a central role as it processes the inputs to produce a fire characterization of the landscape and makes it available to other (possible outer) blocks, as detailed below.

In pyO3F, fire spread is modeled through the MTT principle and therefore is based on representing the landscape as a (directed) network with arc weights associated

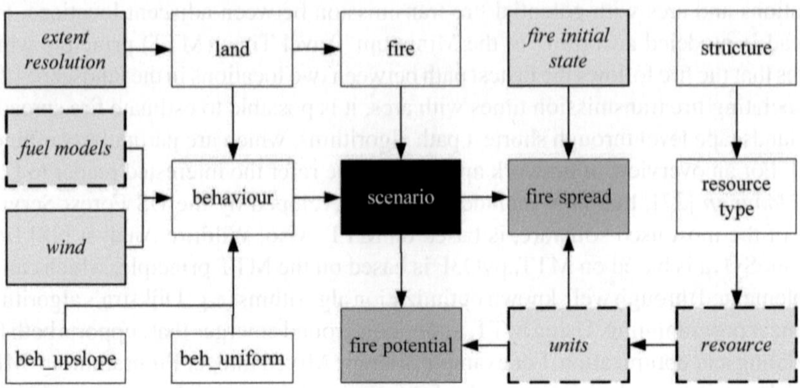

Fig. 1 pyO3F overview

with fire transmission times. The structure of the network depends, first, on the `extent` of the landscape under study and on the `resolution` (i.e. the distance between any two nodes) and, second, on the burnable areas of the landscape - defined by classes `land` and `fire`.

The fire transmission time of each arc depends on the slope, fuels and wind, and is obtained by the class `behaviour`. The slope is calculated from the altitudes of the landscape (owned by `land`). Fuels are characterized by fuel models that, by default, are derived from land use information also owned by `land` but can also be an input. Wind is an input and can be of different types according to the implementations of the base class `behaviour` (currently, `beh_upslope` and `beh_uniform`).

The class `scenario` uses the information from other classes to define an MTT network (that is, a network representing the landscape with estimates of the fire transmission times associated with each arc). Different wind or fuels correspond to different scenarios.

Firefighting resource's activities are represented by a network that differs from the fire network. Resources and fire networks share the same landscape, but they do not share nodes or arcs. The nodes of a resource network correspond to locations where a resource may be positioned (e.g., for suppression or detection), and the arcs correspond to possible movements between two adjacent locations. The class `structure` owns the base network for the resources that move in the same way: roads for terrestrial resources and a grid for aerial resources. Resource types may share the structure but have different attributes, e.g. different speeds on the same type of roads. Class `resource` includes the information of the structure and resource type and also, possibly, the initial position of the resource. The class `units` allow to detail the activity of each particular unit of a resource. For example, two units of the same resource may follow different routes in the landscape and attack fire in different locations.

The interaction between a resource network and the fire network is modeled through proximity. For a type of resource activity (e.g., suppression), an interaction radius is defined. If the distance of a resource node to a fire node is less than or equal to this radius, the resource may interact with the fire (e.g. suppress fire).

In the next section, we detail the core classes of pyO3F, which are responsible for setting scenarios and resources. The classes `fire spread` and `fire potential` are discussed in a continuation paper [1].

4 Core Classes

Core classes are responsible to establish the ground to fire spread simulations, fire potential estimates, and custom fire-related optimization. These include allowing the definition of scenarios and resources. In this section, we detail how that is achieved.

4.1 Land and Fire

In order to use the framework an object of a class `O3F` must be created. In the constructor of this class other classes are created. The class `land` is the first class created by O3F and is responsible for keeping the extent and resolution. The extent of the landscape (i.e., the rectangle corresponding to the area under study) can be defined through a bounding box defined by its coordinates or through a shape file.

We will exemplify the use of pyO3F with the Portuguese municipality of Baião and surroundings considering `extent=(1468, 157830, 21563 and 175535)` (EPSG:3763 - ETRS89 / Portugal TM06). The corresponding rectangle has approx. 20km by 18km. pyO3F is based on representing the landscape by a hexagonal tiling. The user can choose the accuracy of this representation when setting the distance between adjacent pairs of hexagons centers, i.e. the *resolution*. This choice must be a balance between the extent, the accuracy of the representation, and the expected computational effort of the approaches to be used.

Based on the landscape and on the given resolution, a hexagonal grid graph is constructed where each node is associated with an hexagon and has, as an attribute, the coordinates of its center.

Information about land use and altitudes is obtained from public sources [29]. Land use, at a first level, divides the landscape into polygons belonging to one of nine classes: (1) artificialized territories, (2) agriculture, (3) pastures, (4) agroforestry areas, (5) forests, (6) bushes, (7) open or sparsely vegetated areas, (8) wetlands, and (9) surface water bodies. Levels two, three, and four further detail land use. For example, class 5.1.1.5 (level 4) corresponds to a forest of eucalyptus. For each node, the predominant land uses of level 1 and 4 of its hexagon are kept.

Figure 2 displays the land use level 1 and altitudes for Baião. In Fig. 3, the hexagons corresponding to the landscape are displayed for two resolutions: 1000 meters and 250 meters.

The main purpose of the class `fire` is to maintain a network that characterizes the landscape from the point of view of fire. This network is obtained from the *land* network by removing unburnable nodes (derived from their land uses, e.g. nodes associated with surface water bodies will be removed), except for nodes that correspond to sensible locations that may play an important role in some applications. An arc of the *fire* network represents the possible transmission of fire between adjacent burnable nodes or from a burnable node to a sensible node.

4.2 Scenario and Related Classes

4.2.1 Overview

A `scenario` object is built on the fire network and owns an MTT network, i.e. a network where each node represents a location and each arc has a fire transmission

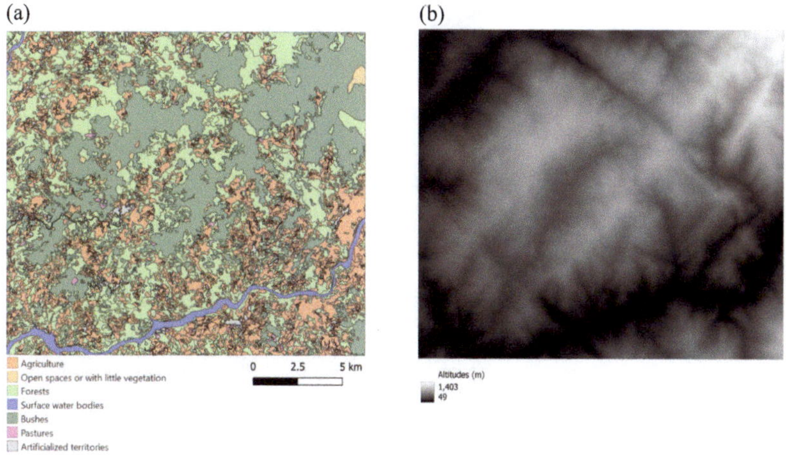

Fig. 2 Land use and altitudes of the Baião extent. **a** Land use level 1. **b** Altitudes

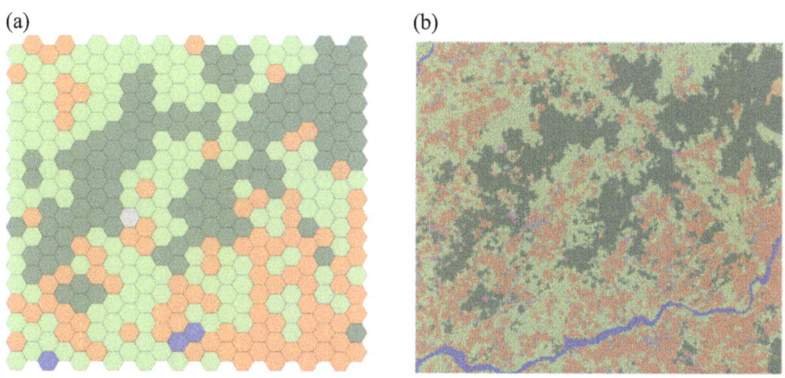

Fig. 3 Representing the landscape with hexagonal cells. **a** Resolution of 1000 m. **b** Resolution of 100 m

time as attribute. An example is provided in Fig. 4 where, for the sake of clarity, a gray scale represents the fire transmission times.

We now describe the existing alternatives to obtain the fire transmission times in pyO3F.

Fire transmission times (or, equivalently, ROS) depend on slope, wind, and fuels. The slope is obtained from the altitudes. Wind is an input. A crucial issue is characterizing the fuel. We use the fuel models from [30] that were conceived for Portugal. Fuels can be derived from land uses or can be inputted directly (through a shapefile or a function). When derived from land use, a correspondence is made between each category of land use and the fuel models for Portugal [30, 31].

The class `behavior` is the base class responsible for characterizing the fire behavior of an arc. The only function required to be implemented by a class derived

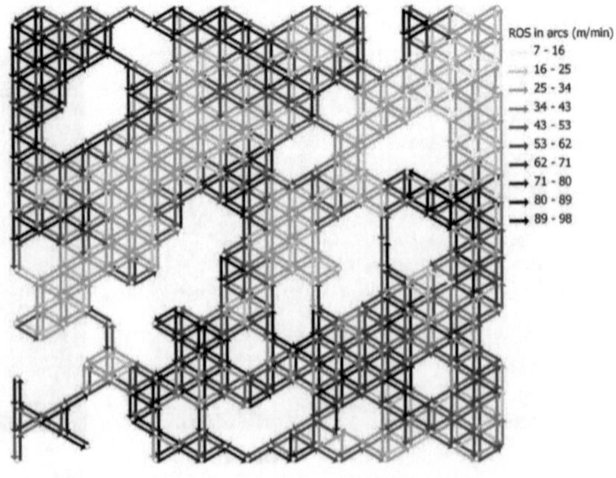

Fig. 4 A scenario with the ROS of all arcs. Upslope wind at 30 km/h

from `behavior` is the `arc_ros` function, which returns the fire ROS using the three-dimensional coordinates, the fuel models of the origin and destination nodes, and the type of wind. Optional functions may be used to estimate the flame length and the fire intensity.

4.2.2 Types of Winds

The current version of pyO3F considers two main methods to estimate the ROS: when the wind is *upslope* and when the wind is *uniform*. An example is provided in Fig. 5 and an explanation follows.

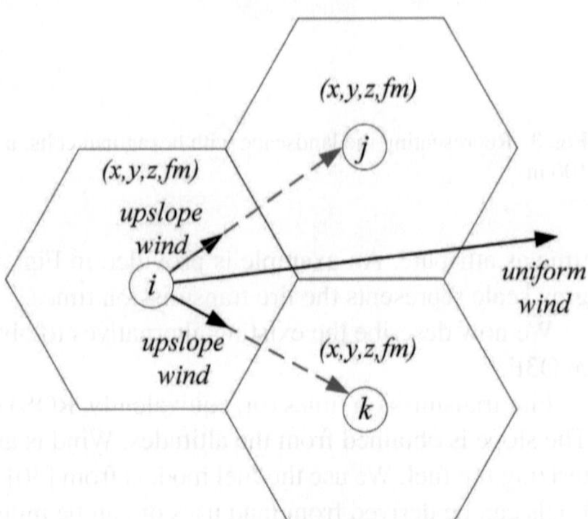

Fig. 5 Types of winds: upslope and uniform. Each node has the three dimensional coordinates and a fuel model (*fm*)

4.2.3 Upslope wind

In an upslope type of wind, the direction of the wind is aligned with each arc. The upslope wind leads to the highest possible ROS of the arc (for the same fuels and, of course, slope). Note that this type of wind does not occur in practise as it implies that more than one wind vector exists at the nodes and therefore leads to a scenario worst than any actual one with the same maximum wind speed.

In pyO3F, currently, there are two main alternatives to estimate the ROS of an arc with upslope wind: Rothermel's and double rule.

Rothermel's fire behavior model

The widely used Rothermel's fire behaviour model [32] estimates the ROS based on the fuel, slope and wind according to:

$$R = R_0(1 + \Phi_w + \Phi_s), \tag{1}$$

where R_0 is the *no slope no wind* ROS (that is influenced by the fuel) and Φ_w and Φ_s are wind and slope factors, respectively.

Rothermel's model has more than ten input parameters (some related to fuel particles, others to fuel array, and some others to the environment) and several constants derived experimentally. A detailed description of the model is provided in [33]. We used the computational implementation of the Rothermel model of Behave Plus 6 software [34] to estimate ROS for each of the 18 fuel models defined in [30, 31] for Portugal and the combinations of 400 slope values and 64 wind intensities. In this way, the ROS given by the model of Rothermel is kept in a set of dictionaries with keys fuel model, wind and slope.

Characterizing fuels and/or estimating ROS are research topics on their own. We refer the interested reader to [35], chapters 6 and 7. Rothermel's model is not suitable for negative slopes. For negative slopes, the following method is used (which works for any slope).

Double rule

One long-held rule of thumb, as mentioned in [12], is to consider that ROS doubles for every 10 degrees increase in slope. Assuming a similar rule for decreasing slopes, results in:

$$R = R_0(w)2^{(s/10)}, \tag{2}$$

where $R_0(w)$ is the ROS with no slope and a wind intensity of w which is estimated by Behave Plus 6 and s is the slope in degrees.

4.2.4 Uniform Wind

In a uniform type of wind, the direction of the wind may not be aligned with the arcs, as exemplified in Fig. 5, i.e., it may be different from the upslope direction. In this case, Behave 6 is used to obtain the *no slope no wind* ROS (that is, R_0) and the projection method described in [33] is used to obtain the ROS in the desired direction (that is, the arc direction). In this case, the wind direction is common to all arcs, and the ROS on an arc is obtained by projection (assuming the generally accepted elliptical shape for fire spread).

4.2.5 Additional Issues

Arcs with different fuel models. So far, we assumed that the fuel models associated with the origin and destination nodes of an arc were the same. If that is not the case, for example in Fig. 5 $fm_i \neq fm_j$, an average of both ROS is calculated. The increase in resolution (i.e. lower values for the distance between nodes) improves the ROS estimate.

Uncertainty on fuel models. In order to tackle the lack of data / uncertainty on the selection of fuel models, pyO3F allows assigning more than one fuel model to each node. In particular, if no fuel models are inputted, a correspondence between land uses and sets of fuel models is used. For example, we associate land use 5.1.1.1 (forest cork oak) with the set of fuel models 212, 222, 225 and 226. Then, the contribution to the ROS of a given node is defined through a parameter that determines which ROS is selected from the ones obtained from each fuel model: the maximum, minimum, or the average ROS.

4.2.6 Fire Initial State

The class `fire_initial_state` represents the fire status when the fire is detected in a given scenario. The specification of the fire initial status can be provided in two different ways: (i) through a shapefile with polygons representing the burned or burning area, and (ii) through the coordinates of a set of ignitions and the time elapsed until the detection (during which fire spreads).

4.3 Resources

4.3.1 Types of Resources

In pyO3F, the potential positions and movements of a resource are represented in a network different from the fire network. Different types of resources may share

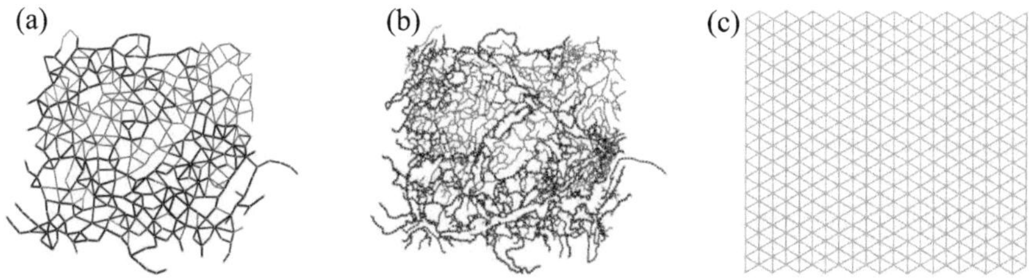

Fig. 6 Base networks for resources. Wider arcs means more important roads/ paths (five categories of roads/paths: from a major road to a path unsuitable for cars). **a** Terrestrial network with contraction parameter 1000 m. **b** Terrestrial network with contraction parameter 250 m. **c** Aerial network with resolution 1000 m

the same network but have different interactions with the fire network and different attributes.

A network can be terrestrial or aerial. Terrestrial networks are derived from *openstreetmap* [36] and converted to a *networkx* graph with the support of *momepy* [37]). Aerial networks are based on a grid (with a given resolution). The resolution of the terrestrial network depends on a parameter that defines the distance below which two nodes are contracted into a single one. Examples of resource networks are provided in Fig. 6.

pyO3F defines six types of resources, the ones shown in Table 1. Each type of resource has a specific interaction with fire that is, in general, based on the proximity of a resource node to the fire nodes. For example, in a *suppression* interaction, a unit of a resource positioned in a node will suppress fire in all fire nodes within a distance less than or equal the parameter `suppression_radius`. The time it takes to suppress fire is `resolution` / `suppression_rate` minutes. In a *safety* interaction, a resource unit can only be at a node if it respects a minimum safe distance, defined by the `safety_radius`, from any fire node. Units of resource types that include suppression or safety interactions may move in their network, in contrast to resources with *vigilance* or *support* interactions that have only one position.

The *support* interaction relates two types of resources, for example a suppression resource (e.g. `ground_direct`) and its base, which is the *support* resource. This interaction provides, for example, the fire nodes that can be reached by the given resource stationed at a specific base.

4.3.2 Actual Resources and Units

In pyO3F, actual resources are classes derived from resource types and inherit the default values of their attributes. A particular relevant issue when deriving a resource from a type is that a subset of nodes of the network can be considered in the resource interactions. For example, for suppression, it is possible to consider only nodes where the flame length and the fire intensity are below given parameters. For movement,

Table 1 Types of resources, their interactions with fire and main attributes

Type of resource	Structure	Interactions	Main attributes
aerial_direct	Aerial	Suppression	suppression_radius, suppression_rate, unit_cost, speed, setup_time
ground_direct	Terrestrial	Suppression, safety	suppression_radius, suppression_rate, unit_cost,
			speed in the five type of roads, setup_time, safety_time, safety_radius
ground_indirect	Terrestrial	Suppression, safety	suppression_radius, suppression_rate, unit_cost, advancement_time
			speed in the five type of roads, setup_time, safety_time, safety_radius
ground_evacuation	Terrestrial	Safety	speed in the five type of roads, setup_time, safety_radius
detection	Terrestrial	Vigilance	vigilance_radius, unit_cost
base	Terrestrial	Support	rsr_served, rsr_interaction, unit_cost, support_time, setup_time

all nodes are considered, enforcing that the resource cannot get closer to fire than a given parameter `safety_radius`.

The network and some attributes of two actual resources are shown in Fig. 7.

Resources represent the potential positions and movements of actual units. Units that move (e.g. a `ground_direct` unit) are characterized by a sequence of nodes

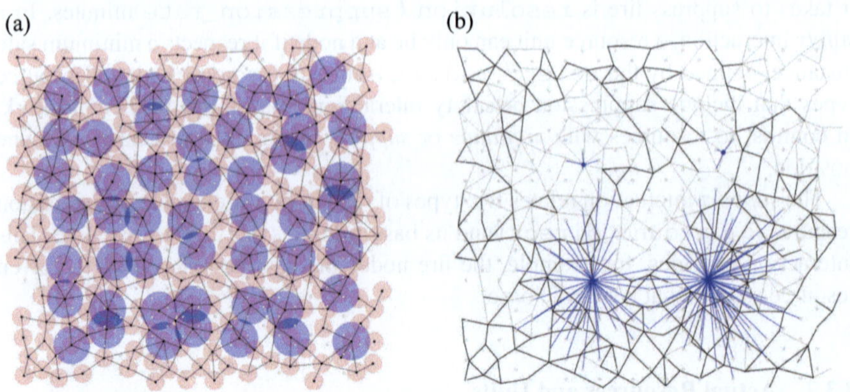

Fig. 7 Examples of the network and some attributes of two resource types. White circles are fire nodes. The resource network is represented with black circles and lines. **a** A `ground_direct` resource type. Blue circles correspond to potential suppression locations and range. Red circle correspond to safety buffers. A resource node cannot be used if there is a fire node inside its red circle. **b** A `base` resource type. Blue lines link a base with a reachable fire node. Bases surrounded by faster roads reach more fire nodes within the available time

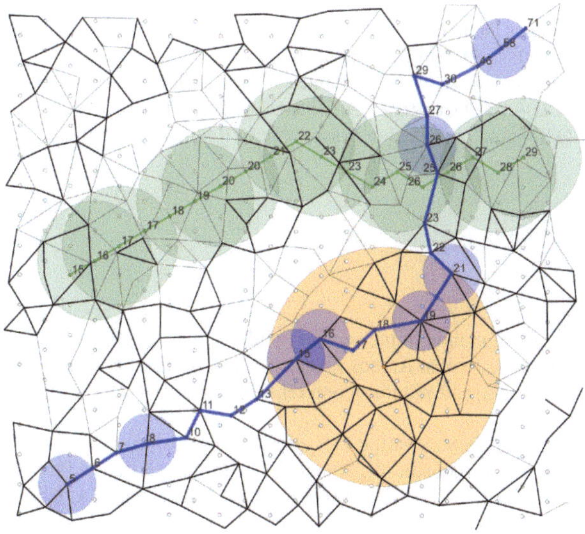

Fig. 8 Examples of three units positions and movements: a `ground_direct` unit in blue, an `aerial_direct` unit in green and a `detection` unit in orange. Arrival times, in minutes, are shown next to the nodes. Circles correspond to suppression (suppression units) or to vigilance (detection unit)

and time instants associated with their arrival and departure times. If they are suppression units, a sequence with the suppression nodes is also kept. If the nodes do not move (e.g. a detection unit), one node defines the position of the resource unit. An example with three units of different resources is provided in Fig. 8.

5 Conclusions

In this paper, we described the architecture and core classes of pyO3F, a python framework that aims to facilitate the implementation of optimization approaches for fire-related problems. pyO3F integrates landscape data (including roads for fire fighting resources), fire behavior models, and network modeling concepts.

A major contribution is the inclusion of resources and their potential interactions with fire.

In a continuation paper [1], we describe the usage of pyO3F for the simulation of fire spread, estimating the fire potential, and implement specific optimization approaches.

Acknowledgements The author acknowledges the collaboration of Marco Marto (improving and testing the `behaviour` class and related ones, correcting bugs, developing a `dash` interface, inputting routes through shapefiles) and David Neto (developing the `beh_uniform` class). The author also acknowledges Paulo Fernandes (Universidade de Trás-os-Montes e Alto Douro) for the files with the fuel models used with BehavePlus 6.

This work has been supported by FCT - Fundação para a Ciência e Tecnologia within the R&D Unit Project Scope UID/00319/Centro ALGORITMI (ALGORITMI/UM) and within project PCIF/GRF/0141/2019 "O3F - An Optimization Framework to Reduce Forest Fire".

References

1. Alvelos, F.: pyO3F - a python framework for wildfire-related optimization, part ii: Usage. In: Alvelos, F., Martins, I., Rocha, A. (eds.) Advances in Optimization and Wildfire: Proceedings of the First Optimization and Wildfire Conference, October 1–4 2024, Luso, Portugal, Springer (2025).
2. Hagberg, A.A., Schult, D.A., Swart, P.J.: Exploring network structure, dynamics, and function using NetworkX. In: Varoquaux, G., Vaught, T., Millman, J. (eds.) Proceedings of the 7th Python in Science Conference (SciPy2008), pp. 11–15. Pasadena, CA, USA (2008)
3. Jordahl, K., Van den Bossche, J. Fleischmann, M., Wasserman, J. McBride, J., G. Jeffrey, Leblanc, F., et al.: Geopandas (Version v0.8.1). Zenodo (2020) http://doi.org/10.5281/zenodo.3946761
4. McKinney, W: Data Structures for Statistical Computing in Python. In Proceedings of the 9th Python in Science Conference (pp. 56–61), Stéfan van der Walt (Ed.) (2010) https://doi.org/10.25080/Majora-92bf1922-00a
5. Harris, C.R., Millman, K.J., van der Walt, S.J., et al.: Array programming with numpy. Nature **585**, 357–362 (2020). https://doi.org/10.1038/s41586-020-2649-2
6. Hunter, J.D.: Matplotlib: A 2D graphics environment. Comput. Sci. Eng. **9**(3), 90–95 (2007)
7. Gillies, S., et al.: Shapely: manipulation and analysis of geometric objects (2007) https://github.com/Toblerity/Shapely
8. Gillies, S., et al.: Rasterio: geospatial raster I/O for Python programmers. Mapbox. (2013) https://github.com/rasterio/rasterio
9. Gurobi Optimization, LLC: Gurobi Optimizer Reference Manual. (2024) https://www.gurobi.com
10. QGIS Development Team: QGIS Geographic Information System. QGIS Association (2025) https://www.qgis.org
11. Sullivan, A.L.: Wildland surface fire spread modelling, 1990–2007. 1: Physical and quasi-physical models. Int. J. Wildland Fire **18**(4), 349–368 (2009)
12. Sullivan, A.L.: Wildland surface fire spread modelling, 1990–2007. 2: Empirical and quasi-empirical models. Int. J. Wildland Fire **18**(4), 369–386 (2009)
13. Sullivan, A.L.: Wildland surface fire spread modelling, 1990–2007. 3: Simulation and mathematical analogue models. Int. J. Wildland Fire **18**(4), 387–403 (2009)
14. Papadopoulos, G.D., Pavlidou, F.-N.: A comparative review on wildfire simulators. IEEE Syst. J. **5**(2), 233–243 (2011)
15. Finney, M.A.: FARSITE, Fire Area Simulator–model development and evaluation. US Department of Agriculture, Forest Service, Rocky Mountain Research Station (1998)
16. Ghodrat, M., Shakeriaski, F., Fanaee, S.A., Simeoni, A.: Software-based simulations of wildfire spread and wind-fire interaction. Fire **6**(1), 12 (2022)
17. Cardil, A., Monedero, S., Schag, G., de-Miguel, S., Tapia, M., Stoof, C.R., Silva, C.A., Mohan, M., Cardil, A., Ramirez, J.: Fire behavior modeling for operational decision-making. Curr. Opin. Environ. Sci. & Health **23** (2021)
18. Tymstra, C., Bryce, R.W., Wotton, B.M., Taylor, S.W., Armitage, O.B.: Development and structure of Prometheus: the Canadian wildland fire growth simulation model. Natural Resources Canada, Canadian Forest Service, Northern Forestry Centre, Information Report NOR-X-417 (Edmonton, AB) (2010)

19. Alexandridis, A., Russo, L., Vakalis, D., Bafas, G., Siettos, C.: Wildland fire spread modelling using cellular automata: evolution in large-scale spatially heterogeneous environments under fire suppression tactics. Int. J. Wildland Fire **20**(5), 633–647 (2011)
20. Zheng, Z., Huang, W., Li, S., Zeng, Y.: Forest fire spread simulating model using cellular automaton with extreme learning machine. Ecol. Model. **348**, 33–43 (2017)
21. Pais, C., Carrasco, J., Martell, D.L., Weintraub, A., Woodruff, D.L.: Cell 2fire: a cell-based forest fire growth model to support strategic landscape management planning. Front. For. Glob. Chang. **4**, 692706 (2021)
22. Gonzalez-Olabarria, J.R., Carrasco, J., Pais, C., Garcia-Gonzalo, J., Palacios-Meneses, D., Mahaluf-Recasens, R., Porkhum, O., Weintraub, A.: A fire spread simulator to support tactical management decisions for mediterranean landscapes. Front. For. Glob. Chang. **6**, 1071484 (2023)
23. Trucchia, A., D'Andrea, M., Baghino, F., Fiorucci, P., Ferraris, L., Negro, D., Severino, M., et al.: PROPAGATOR: an operational cellular-automata based wildfire simulator. Fire **3**(3), 26 (2020)
24. Finney, M.A.: Fire growth using minimum travel time methods. Can. J. For. Res. **32**(8), 1420–1424 (2002). https://doi.org/10.1139/x02-068
25. Ahuja, R. K., Magnanti, T. L., Orlin, J. B.: Network Flows: Theory, Algorithms, and Applications. Prentice hall (1993)
26. Aparício, B.A., Pereira, J.M., Santos, F.C., Bruni, C., Sá, A.C.: Combining wildfire behaviour simulations and network analysis to support wildfire management: a mediterranean landscape case study. Ecol. Ind. **137**, 108726 (2022)
27. Finney, M.A.: An overview of FlamMap fire modeling capabilities (2006). In: Andrews, P.L., Butler, B.W., Fuels Management-How to Measure Success: Conference Proceedings. 28-30 March 2006; Portland, OR. Proceedings RMRS-P-41, pp. 213--220. Fort Collins, CO: US Department of Agriculture, Forest Service, Rocky Mountain Research Station (2006)
28. Ramírez, J., Monedero, S., Buckley, D.: New approaches in fire simulations analysis with Wildfire Analyst. In: Proceedings of the 5th International Wildland Fire Conference, pp. 9--13. Sun City, South Africa (2011)
29. Direção-Geral do Território: Carta de Uso e Ocupação do Solo - 2018 (2022). https://geo2.dgterritorio.gov.pt/cos/COS2018/COS2018v2-shp.zip
30. Fernandes, P., Gonçalves, H., Loureiro, C., Fernandes, M., Costa, T., Cruz, M.G., Botelho, H.: Modelos de combustível florestal para Portugal. Actas Do 6o Congresso Florestal Nacional, pp. 348--354. SPCF, Lisboa, Portugal (2009)
31. Fernandes, P., Loureiro, C.: Modelos de combustível florestal para Portugal. Documento de referência, Technical report, Departamento de Ciências Florestais e Arquitetura Paisagista, UTAD, Portugal (2021)
32. Rothermel, R.C.: A mathematical model for predicting fire spread in wildland fuels (Vol. 115). Intermountain Forest and Range Experiment Station, Forest Service, USA (1972)
33. Andrews, P.: The Rothermel surface fire spread model and associated developments: A comprehensive explanation. USDA Forest Service, Rocky Mountain Research Station [Techreport]. General Technical Report RMRSGTR-371.(Fort Collins, CO) (2018)
34. Andrews, P.: Current status and future needs of the behaveplus fire modeling system. Int. J. Wildland Fire **23**(1), 21–33 (2013)
35. Rego, F.C., Morgan, P., Fernandes, P., Hoffman, C.: Fire Science. Springer (2021)
36. OpenStreetMap contributors: OpenStreetMap [Data set]. OpenStreetMap Foundation. Available as open data under the Open Data Commons Open Database License (ODbL) at openstreetmap.org 36 (2024)
37. Fleischmann, M.: Momepy: urban morphology measuring toolkit. J. Open Source Softw. **4**(43), 1807 (2019). https://doi.org/10.21105/joss.01807

pyO3F—A Python Framework for Wildfire-Related Optimization Part II: Usage

Filipe Alvelos

Abstract pyO3F is a Python framework to support the implementation of optimization approaches to wildfire-related problems. This paper follows a previous one (pyO3F—A python framework for wildfire-related optimization, Part I: Design and fundamentals) in which the design and the core classes of the Python framework pyO3F were introduced, along with a discussion of related work. Here, we describe and exemplify how pyO3F can be used to address wildfire optimization problems.

Keywords Software · Fire spread simulation · Wildfire supression · Mixed integer programming

1 Overview

pyO3F supports the implementation of optimization-based approaches by defining fire scenarios. The landscape under study is represented as a network with nodes associated with cells and arcs representing the potential transmission of fire between two adjacent cells. A scenario is characterized by the fire transmission times resulting from fire behavior models that use as input slopes, fuels, and winds. The class `scenario` may be used directly in optimization approaches as a source of information on the landscape and data related to fire. In a base use, the only information required to define a scenario is the type of wind and its intensity. It can also be used by classes `fire_potential` and `fire_spread` that estimate the fire potential and simulate the fire spread. These two classes differ in the presence (i.e at least one ignition has occurred) or absence of fire. While `fire_potential` is used to obtain measures of the potential impact of fire in the cells or regions of the landscape in a scenario with no active fires, `fire_spread` is used to simulate fire spread from a given initial fire state (e.g. an initial fire perimeter). In both classes, it is possible to include the presence of firefighting resources positioned by the user (or by some

F. Alvelos (✉)
Department of Production and Systems/ALGORITMI Research Center/LASI, School of Engineering, University of Minho, Braga, Portugal
e-mail: falvelos@dps.uminho.pt

F. Alvelos et al. (eds.), *Advances in Optimization and Wildfire*, Lecture Notes in Operations Research, https://doi.org/10.1007/978-3-032-03108-2_4

optimization procedure). Classes `fire_potential` and `fire_spread` were thought to be used in the two stages of fire management where optimization has been used mainly: prevention (e.g. optimization of fuel treatments, surveyed in [1]) and suppression (surveyed in [2] and [3].

Two examples of problems in which pyO3F was used are:

- Dispatching, positioning and routing resources for wildfire initial attack [4];
- Deciding prescriptions and fuel treatments in forest management taking into account fire potentials [5].

The first reference [4] builds on [2] and [6] to model the initial attack by multiple resources that move in networks representing actual ones (e.g., roads) and that are different from the fire network. Furthermore, resources avoid fire on their routes to the attack positions and have different suppression radius according to their type / skill / dimension. Resources are available after a preparation time and can only attack fire in positions reached before fire and for which there is a route that does not cross fire at any instant (fire is avoided). pyO3F is used directly. A `scenario` and a set of resource objects provide data to a MIP model. The approach, which iterates between MIP and shortest paths, also uses pyO3F to determine safe routes to the resources (paths that avoid fire). Class `fire_spread` is used to visualize the solution obtained.

In the second reference [5], class `fire_potential` is used to obtain fire paths with a rate of spread above a given threshold (termed unacceptable paths) for scenarios with fuels associated with different forest management alternatives. In [7], this approach was extended to integrate decisions to the use of firefighting resources.

pyO3F have been used in other problems: prepositioning firefighting resources in a MIP integrating several scenarios to deal with uncertainty [8], location problems with network design [9, 10], and without [11], and a problem of finding disjoint paths avoiding fire (for safe movements of resources and evacuation of people) [12].

In Sect. 2, we first describe how a scenario is created. The use of the `fire_spread` and the `fire_potential` classes is described and exemplified in Sects. 3 and 4, respectively. In Sect. 5, the direct use of the class `scenario` is exemplified with a MIP model for initial attack problem. Section 6 is devoted to how pyO3F can support the implementation of metaheuristics. In the last Section, Sect. 7, we derive the main conclusions.

2 Creating a Fire Scenario

Any use of the pyO3F begins by creating an `O3F` object with arguments defining the landscape `extent` through a bounding box defined by its coordinates and the `resolution` to be used in its discretization. For example, the creation of a `O3F` object, declared as `of`, for the Portuguese Baião Municipality and a resolution of 1000 m the instruction is:

`of=o3f.O3F((1468, 157830, 21563, 175535), 1000).`

Data from land use and altitudes corresponding to the extent are extracted in a transparent way to the user.

A `scenario` object can be used by optimization approaches directly, or defined to simulate fire spread or to estimate the fire potential, possibly within some optimization procedure. A `scenario` receives as input the wind and, if other than the default values (derived from the land use, see [13]) are desired a set of fuel models, one for each node. The fire transmission times are obtained with a fire behavior model. An example of the creation of a `scenario` object with an upslope wind with a speed of 30km/h and default fuels is:

`scen=of.create_scenario(wind=('upslope',30), fuel_models=None).`

The wind can be of different types (described in [13]). Configuration parameters, defined in a configuration file along with many others not mentioned for simplicity, allow to define the land use level to be used in the default fuels and how land uses are converted to fuels.

An object of class `scenario` owns a *networkx digraph* object, named `G`, that represents the MTT network. In addition to the attributes provided by the *networkx digraph*, each node `i` has its absolute coordinates `scen.G.nodes[i]['coord_abs']` as an attribute and each arc `(i,j)` has a fire transmission time `scen.G.edges[i,j]['trans_time']` attribute.

3 Fire Spread Simulation

Fire spread simulation amounts to, given a fire initial state, estimate the fire arrival time at each node of the fire and the corresponding fire paths. pyO3F can perform a simulation of fire spread in the presence or absence of resources.

In order to simulate fire spread, a `fire_initial_state` object is created. In addition to a `scenario`, at least one ignition (defined by the ID of the nodes, previously identified positions, or node coordinates - absolute or relative), and the instant when the fire is detected (or, replacing the latter two, a shapefile defining the burned area when the fire(s) is detected). An instruction for the definition of a fire initial state is:

`fire_ini=of.create_fire_initial_state(scen, ignitions, alert_time),`

where `ignitions` is a set of ignitions that initiate fire spread and `alert_time` is the instant when the fire is detected. At that instant, the attributes of `fire_ini` `ignited`, `burned`, `burning`, `active_arcs` of `fire_ini` have lists with identifiers of the ignition nodes, nodes burned in the time interval between the ignition and alert time (no longer transmit fire), nodes burning at the alert time (these are the perimeter nodes, fire reached them and they have at least one arc to an unburned node) and arcs from a burning node to an unburned node, respectively.

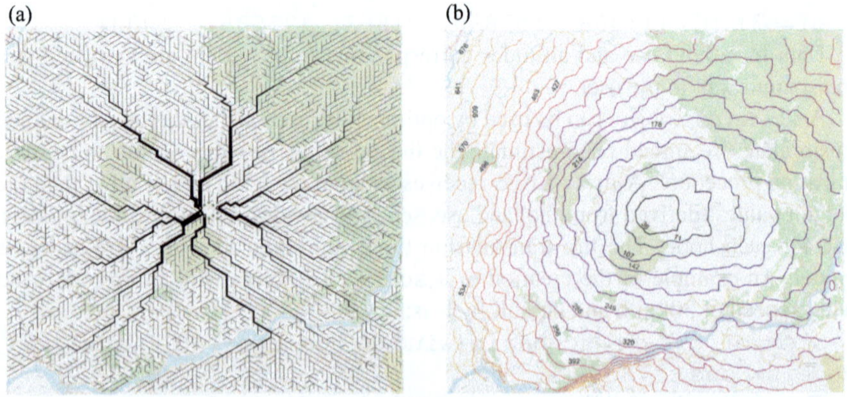

Fig. 1 Two fire spread simulation outputs. **a** Fire paths. Wider arcs belong to more fire paths. **b** Isochronous of fire arrival times

3.1 *Without Resources*

Fire spread simulation without resources is obtained with the instruction `spr=of.run_spread(fire_ini, units=None)`

Object `spr` has all the outputs of the simulation including a *networkx digraph* object, named `G`, with all the fire paths and the fire arrival times at each node (`spr.G.nodes[i]['fire_arr_t']`).

Through `of.spread_to_shp(spr, to_shp_scen=True, to_shp_ini=True)` it is possible to create shapefiles with the spread graph, and, optionally, the input scenario and fire initial state. Figure 1 illustrates two different fire spread results.

Additional outputs include, at given time instants: burned area, fire perimeter, number of active arcs, maximum fire rate of spread (ROS) in an active arc, out degree of a burning node.

This information is shown in graphs as the one in Fig. 2.

3.2 *With Resources*

Fire spread can take into account the presence of resources, which should not be confused with any optimization. Optimization will be addressed in Sects. 5 and 6. The simulation receives the position of each (suppression) resource unit and takes its effect in the fire spread into account by forbidding fire transmission from any fire node within a given (suppression) radius centered at the unit resource position.

As an example, an actual resource named `crew` derived from the base resource `ground_direct` (the default direct attack resource, owned by class O3F) is created with `of.add_resource('crew','ground_direct',None)`

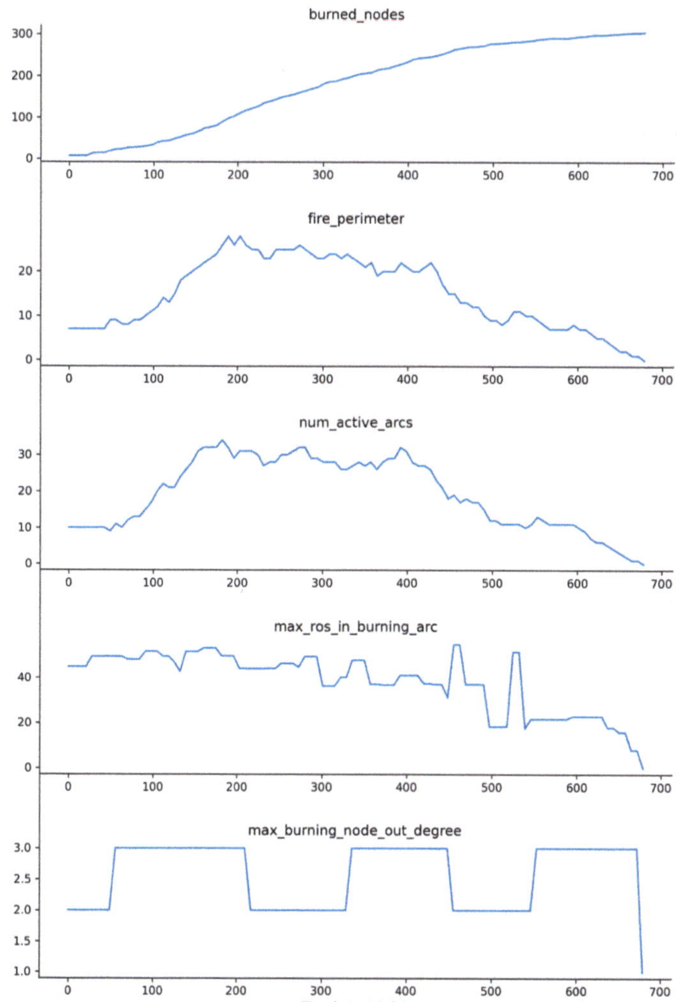

Fig. 2 Spread graphs

The third argument specifies the initial position of the resource. Optional arguments are used, for example, to define the attack positions as subset of the nodes of the resource network.

A `crew` unit with index 0 is created through

`unit=of.create_resource_unit(0, 'crew')`

Two lists are used to indicate the route of the unit (that is, the sequence of nodes the unit visits) and a list of suppression nodes, here named `seq` and `seq_sup`. Other optional arguments can be given to, for example, specify the arrival and departure times of the unit at each node. If those times are not provided, they will be calculated assuming that there are no waiting times on the route. The nodes in the lists may come from previously identified positions, coordinates (absolute or relative), or a shapefile. In the latter two, the nearest node of the given location is selected.

The following instructions allow to define and visualize a route of a resource unit through the nodes in the list `seq` and the suppression nodes in the list `seq_sup`). They also obtain and visualize the corresponding fire spread:

```
units=[unit.set_route_by_seq(seq, seq_sup_nodes=seq_sup)]
spr=of.run_spread(fire_ini, units=units)
of.spread_to_shp(spr,to_shp_ini=True, to_shp_scen=False)
of.units_to_shp(units,agg_by_rsr=True)
```

Figure 3 illustrates the effect of the presence of resources in the fire spread. Each position corresponds to one unit.

It is important to note that, in these two examples, the position of the resources were chosen arbitrarily and *not* by any optimization procedure in order to illustrate how fire spread can be estimated given the position of a set of resource units. In fact, it can be seen that some units in Fig. 3 could be removed without any change in fire spread, saving effort.

All the measures mentioned for the simulation without resources are also available in the simulation with resources.

pyO3F allows to produce animations with fire spread and resources movement and attack. Some frames of one animation are shown in Fig. 4 (fire is contained) and Fig. 5 (fire is not contained).

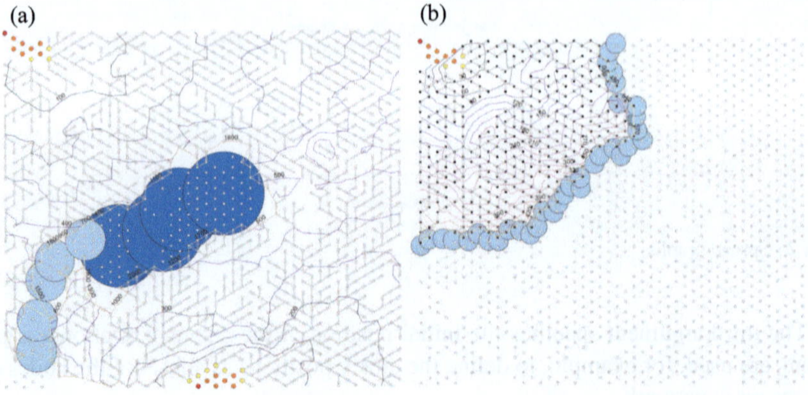

Fig. 3 Two examples of how fire interacts with resources. Red nodes are ignition nodes. Orange nodes are nodes burned at alert time. Yellow nodes are nodes burning at alert time. White nodes inside circles are unburned. Black nodes burned and white did not. **a** Five ground_direct units and four aerial_attack units. With the resources at the given positions, the two fires cannot be contained. **b** One fire is contained

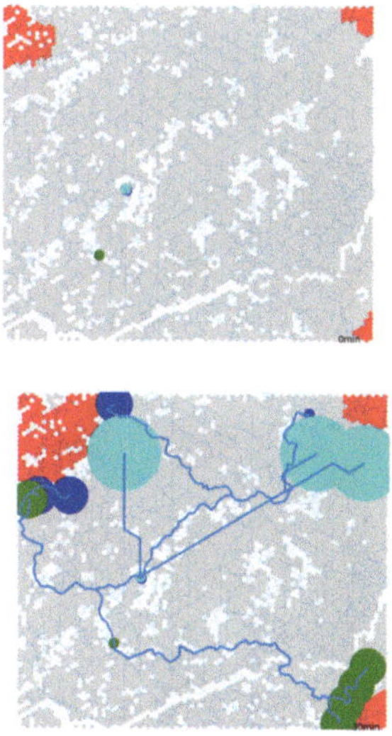

Fig. 4 Frames from an animation with (a contained) fire spread and resources movement and attack. Instants: 0, 10, 20, 30, 40, and 50 minutes. Fire cells are red, other colors correspond to resources (small circles for movement, large circles for attack, lines for route)

4 Fire Potential Estimates

Fire potential refers to estimates of measures for a given scenario considering all potential ignitions in the landscape independently, i.e, one at a time. Two types of measures are considered: (i) based on the initial fire spread, i.e. the fire spread until a given evaluation instant and (ii) based in the paths fire will follow to all nodes. Additionally, paths with a ROS higher than a given threshold, termed unacceptable paths, can be identified. In all cases, resources may or may not be present.

Examples of the type of measures based on the initial fire spread is the proportion of ignitions a node burned until the evaluation instant (i.e., the burn probability) and the sum of the ROS of the outgoing burning arcs also at the evaluation instant.

Measures based on fire paths may be relative to arcs or nodes. An example of a measure relevant to an arc is the number of fire paths it belongs to; for nodes, examples are the sum of the ROS of its outgoing arcs, the out degree and the betweenness centrality. Figure 6 exemplifies some of those measures without resources and with one resource.

An example of direct use of the class `fire_potential` with resources is in location problems. In these type of problems, in general, it is intended to maximize the coverage of "demands" by "facilities". In the context of wildfire, "demands" may be measures provided by class `fire_potential` mentioned above and the "facilities" are resources.

Fig. 5 Frames from an animation with (an uncontained) fire spread and resources movement and attack. Instants: 0, 42, 84, 126, 168, 210, 252, and 294 minutes. Fire cells are red, other colors correspond to resources (small circles for movement, large circles for attack, lines for route

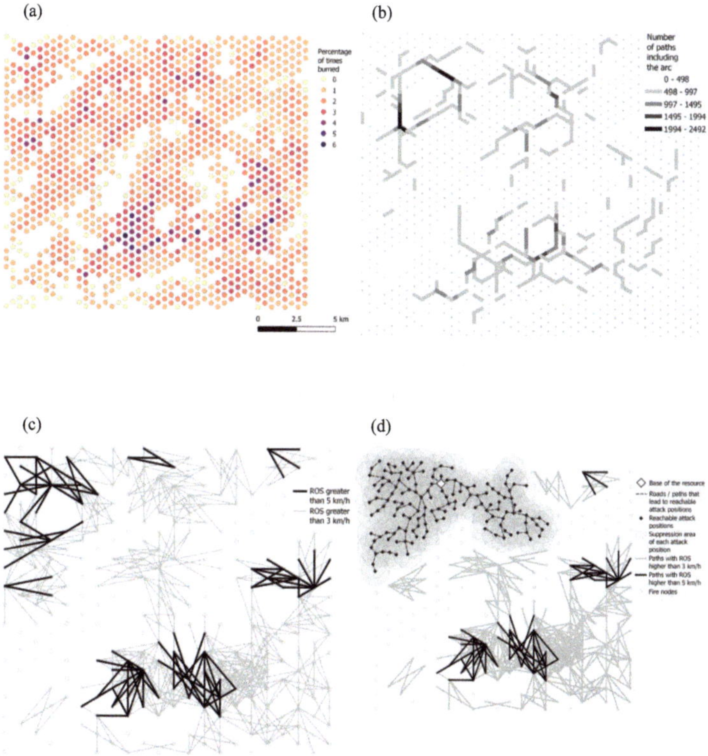

Fig. 6 Examples of fire potential estimates. **a** Based on initial fire spread: proportion of ignitions a node burned until an evaluation time (30 minutes). **b** Based on fire paths: number of paths that include each arc. **c** Unacceptable paths: the average ROS of all paths with a length between 3000m and 4000m is obtained and compared against the threshold. **d** Unacceptable paths with a resource. A resource unit based at the signaled location can reach locations within 15 minutes from the base and has a suppression radius of 1000m

5 Direct Use of `Scenario`

We provide an example of the direct use of class `scenario` with a MIP model for an initial attack problem. This problem is a simplification of the one addressed in [4]. In that reference, a significant effort was made to model more accurately the positioning of resources used so far ([2, 6] and references there to previous works) by using different networks for different resources and model the interaction between fire and resources through safety and suppression relations. In this example, we only consider the suppression relation (as in the previous works mentioned), that is, the interaction between fire and terrestrial resources during the route to an attack position is not considered as it is in [4]. The problem consists in, after a fire is detected, positioning a set of units of a set of resources with lexicographic objectives: first, minimize the burned area after a given time period, second, minimize the cost associated with the resources used, and third, minimize the sum of their arrival times at the attack positions.

For clarity, we introduce the model in mathematical terms, pointing out the data and parameters provided by the class `scenario`. We note that there is a correspondence between the notation used and the attributes of `scenario`. For example, `G.edges[i,j]['trans_time']` is the same as m^f_{ij}, both correspond to the fire transmission time between nodes i and j of the fire network.

We now introduce the notation. The set of resources is represented by R and indexed by r. Entities and parameters related to fire have superscript f. The `python` implementation provides the following classes:

- `G_f`, `networkx` object representing the fire network;
- `i_state`, an object representing the fire initial state;
- `rsrs`, a dictionary of resource types, `rsrs[rsr]` is a dictionary that characterizes the resource type. For example, `rsrs[rsr]['G']` is the network of resource `rsr`.

The `scenario` class provides the following parameters and data:

- N^f set of nodes of the fire network, `G_f.nodes`;
- I^f set of burning nodes, $I^f \subset N^f$, `i_state.burning_at_alert`;
- A^f set of arcs of the fire network, `G_f.edges`;
- m^f_{ij} fire transmission time between adjacent nodes i and j, $ij \in A^f$, `G_f.edges[i,j]['trans_time']`;
- N^r set of nodes of the resource type r, $r \in R$, `rsrs[rsr]['G'].nodes`;
- A^r set of arcs of the network of resource type $r, r \in R$, `rsrs[rsr]['G'].edges`;
- m^r_{pq} movement time between adjacent nodes p and q for resource type r, $r \in R$, $pq \in A^r$, `rsrs[rsr]['G'].edges[(i,j)]['move_time']`;
- L^{rf} set of pairs (k, i), where k is a resource node of the network r and i is a fire node, such that the distance between k and i is less than the suppression radius of r, in the pyO3F implementation, lists are used: `rsrs[rsr]['sup_point_fire_circle_rsr'][i]` is the list of nodes k of resource `rsr` with a distance to fire node i less than its suppression radius.

The user defines the available resources and the inputs:

- b^r number of available resources of type r, $r \in R$, `num_units[rsr]`;
- c^r unit cost of a resource of type r, $r \in R$, `rsrs[rsr]['unit_cost']`;
- l^r base node for type of resource r, $r \in R$, $l^r \in N^r$, `rsrs[rsr]['ini_pos_node']`.

It also defines the instant when the burned area is evaluated, T.

Using a `O3F` function, the user calculates, for each resource, the shortest path tree with l^r as root using m^r_{pq}, obtaining:

- h^r_k - duration of the quickest path of a resource of type r from its base l^r to a node k, `quickest_arrival_time[rsr][k]`.

Fire spread is modeled with the following decision variables:

$$t_i^f - \text{instant of the fire arrival at node } i, \qquad i \in N^f.$$

As an example, the declaration of these variables, is:
`t_f = self.model.addVars(G_f.nodes, ub=G_f.fire_arrival_upper_bound, vtype=gp.GRB.CONTINUOUS)` where `G_f.fire_arrival_upper_bound` is an upper bound on the fire arrival time at the nodes.

The following decision variables are used to express the burned area at instant T.

$$y_i^f = \begin{cases} 1, & \text{if node } i \text{ burns before instant } T \\ 0, & \text{otherwise} \end{cases}, \qquad i \in N^f.$$

The decisions where (i.e. in which nodes of the resource network) are modeled through:

$$x_k^r = \begin{cases} 1, & \text{if a unit of resource of type } r \text{ attacks fire at node } k \\ 0, & \text{otherwise} \end{cases}, \qquad r \in S, k \in N^r.$$

The model is the following:

$$\text{Lex}$$

$$\text{Min} \sum_{i \in N^f} y_i^f$$

$$\text{Min} \sum_{r \in R, k \in N^r \setminus \{l^r\}} c^r x_k^r \tag{1}$$

$$\text{Min} \sum_{r \in R, k \in N^r \setminus \{l^r\}} h_k^r x_k^r \tag{2}$$

Subject to:

$$t_i^f = 0 \qquad i \in I^f \tag{3}$$

$$t_j^f \le t_i^f + m_{ij}^f + T \sum_{r \in R, ki \in L^{rf}} x_k^r \qquad ij \in A^f \tag{4}$$

$$y_i^f \ge 1 - t_i^f / T \qquad i \in N^f \tag{5}$$

$$\sum_{k \in N^r} x_k^r \le b^r \qquad r \in R \tag{6}$$

$$h_k^r x_k^r \le t_i^f \qquad i \in N^f, r \in R, ki \in L^{rf} \tag{7}$$

$$t_i^f \geq 0 \qquad i \in N^f$$

$$y_i^f \in \{0, 1\} \qquad i \in N^f$$

$$x_k^r \in \{0, 1\} \qquad r \in R, k \in N^r \tag{8}$$

The objective functions, (1), (2), and (3), were already introduced in the problem definition. The constraints (4) and (5) define the perimeter of the fire at the detection time and model fire spread, respectively. Constraints (5) are the fundamental constraints in modeling fire spread with MIP. They state that the fire arrival time at a node is the minimum among the potential fire arrival times from each incoming arc, which implies, when maximizing the fire arrival times, that they correspond to quickest paths. The presence of resources, delays the fire in all the outgoing arcs.

The relation of a node that is burned in T and the arrival of fire in it is established in (6). The number of units available for each resource is limited in (7). Constraints (8) state that a unit can only be placed in a node where the fire did not yet arrive.

An example of an optimal solution for an instance with a single resource, after the translation of the optimal values of the variables to units routes (see above), is provided in Fig. 7.

pyO3F offers support for the definition of MIP objective functions. As long as a convention of names is used (for example, the names used in this example follow that convention, t^f for fire times, y for burned nodes, x for positions), objectives are defined by gurobi linear expressions in a transparent way to the user, and the gurobi solver instructed to use multi-objective optimization with the priorities and weights given (lexicographic order is used, for objectives with the same priority, weights are

Fig. 7 Initial attack

used). The user only has to define a list of 3-tuples in the form `(name, priority, weight)`, where `name` may be defined as, for example, `num_burned` for number of burned nodes, `sum_arrival_times` the sum of the fire arrival times, or, adding `_sensible` to the end of the `name`, the same for sensible nodes (i.e., valuable nodes that must be protected).

6 Support for Metaheuristics

The implementation of metaheuristics in pyO3F is based on the representation of a solution as a set of units of each resource, each with a route (that may have a single node). Depending on the problem, a solution is evaluated through the measures returned by calling `fire_spread` and/or `fire_potential`.

A neighbor of a solution is defined as a solution that results from the change of the position of one unit from the current one to another in a distance between d_{min} and d_{max} where these parameters can be defined in spatial or temporal terms. Intensification and diversification moves are easily implemented through these parameters. For example, an intensification procedure can be local search (setting $d_{min} = 0$ and a small d_{max}, while a diversification procedure can be a jump to a solution with a large values for d_{min} and d_{max}

It is planned to improve the metaheuristics support by:

- implementing base versions for common metaheuristics relying on (randomized) constructive procedures and/or local search (e.g. GRASP and VNS);
- implement other operators (e.g., crossover);
- extend the evaluation of solutions to take into account multiple scenarios to deal with uncertainty;
- define neighborhoods that involve more than one unit.

7 Conclusions

pyO3F is a Python framework for supporting the implementation of optimization approaches for wildfire-related problems. In this paper, preceded by another [13], we described how pyO3F can be used to simulate fire, estimate fire potential, and as a source of data for MIP models or an evaluator and provider of neighborhoods in metaheuristics. We also sketched how the metaheuristics support can be extended. This is a priority for future work as many wildfire problems, namely supression ones, require rapid solutions.

Acknowledgements This work has been supported by FCT—Fundação para a Ciência e Tecnologia within the R&D Unit Project Scope UID/00319/Centro ALGORITMI (ALGORITMI/UM) and within project PCIF/GRF/0141/2019 "O3F—An Optimization Framework to Reduce Forest Fire".

References

1. Chung, W.: Optimizing fuel treatments to reduce wildland fire risk. Curr. For. Rep. **1**, 44–51 (2015)
2. Mendes, A., Alvelos, F.: Iterated local search for the placement of wildland fire suppression resources. Eur. J. Oper. Res. **304**(3), 887–900 (2022). https://doi.org/10.1016/j.ejor.2022.04.037
3. Granda, B., León, J., Vitoriano, B., Hearne, J:. Decision support models and methodologies for fire suppression. Fire **6**(2), 37 (2023)
4. Alvelos, F., Marto, M., Mendes, A.: Dispatching, positioning and routing resources for wildfire initial attack. Netw. **86**(1), 80–102 (2025). https://doi.org/10.1002/net.22278
5. Alvelos, F., Martins, I., Marques, S.: Forest management with fire simulation. Eur. J. Oper. Res. **321**(3), 991–1003 (2025). https://doi.org/10.1016/j.ejor.2024.10.013
6. Alvelos, F.: Mixed integer programming models for fire fighting. In: International Conference on Computational Science and Its Applications, pp. 637-652. Springer, Cham (2018). https://doi.org/10.1007/978-3-319-95165-2_45
7. Alvelos F., Martins I.: Integrating fire suppression in forest management. In: Alvelos, F., Martins, I., Rocha, A. (eds.) Advances in Optimization and Wildfire: Proceedings of the First Optimization and Wildfire Conference, October 1-4 2024, Luso, Portugal, Springer (2025)
8. Marques, F., Agra, A., Alvelos, H. Xambre, A.R., Alvelos, F.: A robust approach for the prepositioning of resources for wildfire suppression. In: Alvelos, F., Martins, I., Rocha, A. (eds.) Advances in Optimization and Wildfire: Proceedings of the First Optimization and Wildfire Conference, October 1-4 2024, Luso, Portugal, Springer. Accepted (2025)
9. Alvelos, F. Neto, D., Cunha, E., Vieira, A. Bento-Gonçalves, A., Moura, S.: Location with network design for wildfire surveillance and response. Presentation at the XXVII EURO Working Group on Locational Analysis Meeting, Aveiro, Portugal, September 14–16, 2022
10. Silva, E., Alvelos, F., Marto, M.: Covering with network design for wildfire promptness. In: Alvelos, F., Martins, I., Rocha, A. (eds.) Advances in Optimization and Wildfire: Proceedings of the First Optimization and Wildfire Conference, October 1-4 2024, Luso, Portugal, Springer (2025)
11. Alvelos, F., Moura, S., Vieira, A., and Bento-Gonçalves, A.: A Bi objective Optimization Approach for Wildfire Detection. In: International Conference on Computational Science and Its Applications, pp. 425 436. Springer, Cham (2023)
12. Pascoal, M., Marto, M., Ahmad, I., Alvelos, F.: The Wildfire Safety Paths Problem, presentation at Optimization and Wildfire Conference, 1–4 october. Luso, Portugal (2025)
13. Alvelos, F.: pyO3F—a python framework for wildfire-related optimization, Part I: design and fundamentals. In: Alvelos, F., Martins, I., Rocha, A. (eds.) Advances in Optimization and Wildfire: Proceedings of the First Optimization and Wildfire Conference, October 1-4 2024, Luso, Portugal, Springer (2025)

A MIP Model for Wildfire Extended Attack

Filipe Alvelos, Marco Marto, and André Bergsten Mendes

Abstract In this paper we address the problem of routing the firefighting resources available for an extended attack to a wildfire. We conceive a mixed integer programming (MIP) model that integrates fire spread and resource operations (movement, direct, and indirect attacks), both spatially and temporally. We consider lexicographic objectives related to fire spread—the protection of sensible points and the minimization of the burned area—and to resources—minimizing the cost of their usage and the duration of their routes. At the core of the model is a set of networks: one for fire and one for each type of resource, linked by covering relations representing suppression and safety. This work extends existing MIP models for fire spread and resources positioning (in particular, extending Alvelos et al. [6]). We report on some preliminary computational experiments to validate the model and assess how difficult it is to solve instances derived from an actual landscape with fire tranmission times obtained by the Rothermel's fire behaviour model.

Keywords Fire spread · Vehicle routing · Mixed Integer Programming · Layered networks

---

F. Alvelos (✉)
Department of Production and Systems/ALGORITMI Research Center/LASI, School of Engineering, University of Minho, Braga, Portugal
e-mail: falvelos@dps.uminho.pt

M. Marto
ALGORITMI Research Center/LASI, School of Engineering, University of Minho, Braga, Portugal

A. B. Mendes
Department of Naval Architecture and Ocean Engineering/RouteLab, University of São Paulo, São Paulo, Brazil
e-mail: andbergs@usp.br

F. Alvelos et al. (eds.), *Advances in Optimization and Wildfire*, Lecture Notes in Operations Research, https://doi.org/10.1007/978-3-032-03108-2_5

1 Introduction

Wildfires have huge impacts on humans and ecosystems. Recent events and studies support the increasing relevance of addressing wildfires. For example, while writing this paper, 27 deaths occurred in the California wildfires of January 2025. The burned area of these fires was 163.74 km^2 and 15,798 structures were destroyed.[1] The frequency of extreme wildfire events more than doubled from 2003 to 2023, and the last seven years included the six most extreme [1].

Operations research has contributed to fire management since the late 1950s, as mentioned in [2]. In recent years, the number of contributions has increased significantly. As a sample, in the Scopus database, there were 221 published papers with the words *wildfire, bushfire, or forest fire* and *operations / operational research or optimization / optimisation* compared to 12 in 2004 and 0 in 1984. With respect to 2020, the number of published papers in 2024 (with the same query) is more than double.

Given the complexity and uncertainty of fire, in many fire-related problems, from prevention to suppression, the gap between fire modeling and optimization remains significant. In the case of fire suppression, we may consider two types of difficulties: modeling the interaction between fire and firefighting resources, and solving the resulting optimization problem(s) efficiently. In this paper, we focus on the first challenge of modelling the problem of how firefighting resources should be safely routed (i.e. keeping a minimum distance to fire) to and among (direct and indirect) attack positions (positions where fire spread is delayed or stopped) in order to minimize the impact of the fire.

We propose a compact mixed integer programming (MIP) model for the problem of attacking a wildfire with a set of firefighting resources. For that purpose, as in previous works (e.g., [3]), we define a network that represents the landscape under study: each node represents a geographical cell, and each arc represents the potential transmission of fire between two adjacent nodes. Each arc has an attribute that corresponds to an estimate of the fire transmission time between the nodes it links. The network is directed since two major factors influencing spread fire rate are wind and slope (along with fuels, as considered in our experiments reported in Section 6).

According to the minimum travel time principle [4], fire spreads by the shortest (fastest) paths. In the problem addressed, there is a set of resources, possibly of different types, that may be used to influence that spread. Ideally, the use of firefighting resources may lead to early fire containment.

Firefighting resources suppress fire through direct or indirect attacks. A network is defined for each resource, where the nodes represent potential attack positions and the arcs represent movements between potential positions. In the nodes of a resource network, other than those where a direct attack is conducted, the resource must keep a given distance from fire, for safety or to work on an indirect attack.

The objective of containing the fire as early as possible is particularly difficult to address, in particular with mixed integer programming (MIP) as shown in [5],

[1] https://en.wikipedia.org/wiki/2025_California_wildfires.

where two MIP models for this objective are conceived. Other objectives, also relevant depending on the landscape under study and, for example, if there is presence of people, are the minimization of the burned area until a certain instant or the maximization of the instant where fire arrives at sensible nodes.

The approach of this paper can be seen as an extension of [6]. In that reference, different networks for fire and resources were already considered, with the implication of modeling the resource's activities and their interaction with fire in a way closer to reality than in previous works. In the same reference, the safety requirement for the (terrestrial) resource paths was also already addressed.

In that reference, the initial attack was considered. The problem was to decide, after a wildfire detection, how many units of each type of resource should be dispatched to the fire, through which route (avoiding fire), and in which position they should attack the fire. A compact MIP and an efficient heuristic were described and tested in an actual landscape.

In the current paper, we address the extended attack problem. This problem has a bigger time horizon, and therefore it is more relevant to consider indirect attacks and that resources may continue their route after the first attack position to other attack positions.

Problems of positioning resources have been addressed since 2000 [3]. In addition to [6], which, as already mentioned, is the basis of this work, other approaches include [7–10]. For additional references, see the literature review of [9] and [11].

We introduce the MIP model by describing its components sequentially. Section 2 is devoted to fire spread, Section 3 to firefighting resources, Section 4 to how fire and resource interactions are modeled and Section 5 to representative objectives that can be addressed with this modeling approach. Section 6 describes the experiments conducted with this model to evaluate its efficiency. In Section 7, we summarize the main conclusions.

2 Fire Spread

As introduced earlier, fire spread is modeled through a network. The following notation, where the superscript f is used to signal a fire set or attribute, is used:

- N^f set of nodes of the fire network;
- A^f set of arcs of the fire network;
- m_{ij}^f fire transmission time between adjacent nodes i and j, $ij \in A^f$;
- I^f ignition node.

Since fire spread follows the quickest paths in the network, the model determines fire arrival times at each node by defining appropriate decision variables:

- t_i^f—instant of the arrival of the fire at node $i, i \in N^f$.

The model is as follows.

$$\text{Max} \sum_{i \in N^f} t_i^f \tag{1}$$

suject to:

$$t_{I^f}^f = 0 \tag{2}$$

$$t_j^f \leq t_i^f + m_{ij}^f \qquad ij \in A^f \tag{3}$$

$$0 \leq t_i^f \leq UB^f \qquad i \in N^f. \tag{4}$$

This model accommodates multiple ignitions straightforward. With some preprocessing, it also accommodates fires that have already spread from the ignition instant to the detection instant. In that case, a fire perimeter, that corresponds to the state of the fire when a first assessment is made, is defined. The nodes in the perimeter are the ones already reached by fire and with at least one outgoing arc to an unburned node and may be seen as ignition nodes. The nodes inside the perimeter (nodes reached by the fire with no outgoing arcs to the unburned nodes) are excluded from the fire network.

The objective function (1) maximizes the sum of the fire arrival times of all nodes. This objective must be interpreted together with constraints (3). These constraints allow the time variables to have values unrelated to shortest paths, for example, they can all value 0. However, if their sum is maximized, the time variable of a node will have the duration of the quickest path from the ignition to that node. Given the objective function and its sense, constraints (3) can be seen as a linearization of $t_j^f = \text{Min}\{t_i^f + m_{ij}^f : i \in \delta^-(j)\}, j \in N^f$, where $\delta^-(j)$ is the set of origin nodes of the arcs incoming at j. Constraints (2) state that the fire is at the ignition node at the initial instant.

The upper bound UB^f of the decision variables ensures that the problem is bounded. If the fire network has strong connectivity, these bounds are not required independently of the ignition node. But if it is not so, nodes unreachable from the ignition (e.g., in other component) may have infinity fire arrival times, turning the problem unbounded, which is avoided by setting this bound on the variables. Preprocessing can also be used to avoid unboundedness by the identification of reachable nodes and, possibly, using also weakly connectivity. A value for UB^f is

$$UB^f = \min\{\sum_{ij \in A^f} m_{ij}^f, |N^f| \max_{ij \in A^f}\{m_{ij}^f\}\}. \tag{5}$$

This model may also be seen as the linear programming dual of the shortest path tree problem rooted at I^f with variables associated with arcs, as detailed in [5].

3 Firefighting Resources

We consider a set of firefighting resources of different types, represented by R and indexed by r. Each *type of resource*, or *resource* for short, is characterized by a network and a set of attributes. Different resources operate on different networks, that is, with varying potential movements and positions as, for example, terrestrial and aerial resources. Different resources may share a network but with different attributes, e.g. different vehicle models have different speeds.

In the description of the model that follows, we do not consider aerial resources which are easier to address than terrestrial ones. A similar approach to the one in [6] for aerial resources, consisting of aggregating their units and removing safety interactions with fire, is straightforward.

3.1 Networks and Routes

The resource networks are represented by the following.

- N^r set of nodes of the layer of resource type r, $r \in R$;
- A^r set of arcs of the layer of resource type r, $r \in R$.

Each resource has a maximum number of attacks, g^r and an initial location, the node where the resource is located when the fire is detected, l^r.

A resource may have one or more units. The set of units of a resource r is represented by U^r and is indexed by u. In a solution, a unit follows a route. In some nodes, it performs attacks.

We now define the decision variables and constraints related to defining a route and its attack nodes of resource unit, relaxing temporal constraints.

$$z^{ru} = \begin{cases} 1, & \text{if unit } u \text{ of resource of type } r \text{ is used} \\ 0, & \text{otherwise} \end{cases}, \quad r \in R, u \in U^r$$

$$x_p^{ru} = \begin{cases} 1, & \text{if unit } u \text{ of resource of type } r \text{ attacks fire at } p \\ 0, & \text{otherwise} \end{cases}, \quad r \in R, u \in U^r, p \in N^r$$

$$v_p^{ru} = \begin{cases} 1, & \text{if unit } u \text{ of resource of type } r \text{ visits node } p \\ 0, & \text{otherwise} \end{cases}, \quad r \in R, u \in U^r, p \in N^r$$

$$w_{pq}^{ru} = \begin{cases} 1, & \text{if unit } u \text{ of resource of type } r \text{ uses arc } pq \\ 0, & \text{otherwise} \end{cases}, \quad r \in R, u \in U^r, pq \in A^r.$$

The constraints are the following:

$$\sum_{q \in \delta^+(l^r)} w_{l^r q}^{ru} = z^{ru} \qquad r \in R, u \in U^r \tag{6}$$

$$\sum_{q\in\delta^+(p)} w_{pq}^{ru} = \sum_{q\in\delta^-(p)} w_{qp}^{ru} \qquad p \in N^r \setminus \{l^r\}, r \in R, u \in U^r \tag{7}$$

$$\sum_{q\in\delta^-(l^r)} w_{ql^r}^{ru} = z^{ru} \qquad r \in R, u \in U^r \tag{8}$$

$$x_p^{ru} \le v_p^{ru} \qquad p \in N^r, r \in R, u \in U^r \tag{9}$$

$$\sum_{p\in N^r} x_p^{ru} \le g^r \qquad r \in R, u \in U^r \tag{10}$$

$$\begin{aligned} v_p^{ru} &= \sum_{q\in\delta^+(p)} w_{pq}^{ru} && p \in N^r, r \in R, u \in U^r \\ w_{pq}^{ru} &\in \{0, 1\} && pq \in A^r, r \in R, u \in U^r \\ z^{ru} &\in \{0, 1\} && r \in R, u \in U^r \\ x_p^{ru} &\in \{0, 1\} && p \in N^r, r \in R, u \in U^r \\ v_p^{ru} &\in \{0, 1\} && p \in N^r, r \in R, u \in U^r. \end{aligned} \tag{11}$$

Constraints (6), (7) and (8) are flow conservation constraints, or state that there is no route if the unit is not used. Constraints (9) do not allow a node to be used in suppression if it does not belong to the route. Constraints (10) limit the number of attack nodes. Constraints (11) are used to identify the nodes of the route through v_p^{ru}.

3.2 Times

Each resource r, $r \in R$, has the following attributes related to time:

- n^r preparation time;
- m_{pq}^r movement time between adjacent nodes p and q, $pq \in A^r$.

We define the following non-negative continuous decision variables for each resource, r, $r \in R$, each unit u, $u \in U^r$:

- t_p^{rua}—instant when the unit u of resource r arrives at node p, $p \in N^r$;
- t_p^{rub}—instant when the unit u of resource r departs from node p, $p \in N^r$.

The following constraints, force the preparation time to be taken into account and relate the two variables of a node and the departure and arrival times of the origin and destination nodes, respectively, of an arc:

$$t_{l^r}^{rub} \ge n^r \qquad r \in R, u \in U^r \tag{12}$$

$$t_q^{rua} \ge t_p^{rub} + m_{pq}^r - M^r(1 - w_{pq}^{ru}) \qquad pq \in A^r, r \in R, u \in U^r \tag{13}$$

$$t_p^{rub} \geq t_p^{rua} - M^r(1 - v_p^{ru}) \qquad p \in N^r \setminus \{l^r\}, r \in R, u \in U^r \tag{14}$$

$$t_{l^r}^{rua} - t_{l^r}^{rub} \leq h^r \qquad r \in R, u \in U^r \tag{15}$$

Constraints (12) state that the units can leave their initial position as soon as the ignition is detected, but after their preparation time. Constraints (13) relate the arrival and departure times of the fire when one node is visited after the other. They also avoid subtours. Constraints (14) relate the arrival time of the fire and the departure time at each node visited.

The constant M^r, is defined in a similar way to UB^f in (5) but with respect to the network of resource r.

4 Linking Fire and Resources

4.1 Delaying Fire and Safety

The constraints that relate fire with attacks are:

$$t_j^f \leq t_i^f + m_{ij}^f + \sum_{r \in R, u \in U^r, p \in F_i^r} t^{fr} x_p^{ru}, ij \in A^f \tag{16}$$

Constraints (16) replace (3) to define the quickest arrival time of fire at the nodes by taking into account potential attacks. The potential fire transmission from a node i to a node j is delayed by t^{fr} by each unit that covers node i. The attack coverage is defined through the sets F_i^r, which contain all the resource nodes that cover a given node i. In the example, of Figure 1, $F_i^r = \{o, q\}$. If a resource attacks at o, ij and ik their potential fire transmission times will increase by t^{fr}. Coverage is defined by an attack radius, a^r, $r \in R$.

A similar relation between fire and resources is defined for safety. A safety coverage is defined through the sets S_i^r, which contain all the resource nodes of a terrestrial network that become unsafe when the fire reaches node i. Similarly, unsafe coverage is defined by a radius, s^r, $r \in R$. Constraints assuring the safety of the resources are:

$$t_p^{rub} \leq t_i^f + M^r(1 - v_p^{ru} + x_p^{ru}) \qquad i \in N^f, r \in R, u \in U^r, p \in S_i^r. \tag{17}$$

In conjunction with (9), constraints (17) state that, in *non attack* nodes, a resource must depart before fire arrives at any fire node they cover. The relations for the attack nodes are modeled in Section 4.2.

An upper bound for the fire arrival time taking into account that resources may delay fire is:

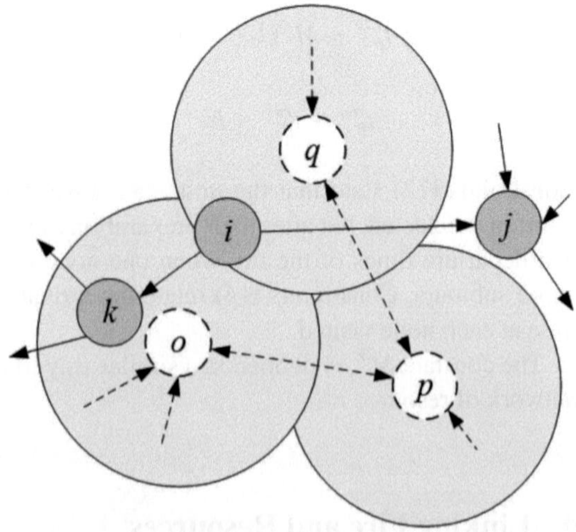

Fig. 1 Modeling fire attack. Dashed nodes and arcs refer to the network of a resource. Solid nodes, filled, in dark grey and arcs refer to the fire network. Large circles refer to the coverage of each resource

$$UB^{fr} = UB^{f} + \sum_{r \in R} |U^r| t^{fr} \tag{18}$$

which replaces UB^f in the domain of the fire variables (4).

4.2 Attacks

We model the two main types of attack: direct and indirect. In a direct attack, fire suppression is attempted by throwing water at the flames. An indirect attack is conducted before the fire arrives and consists of reducing the fuel or its flammability. For example, an indirect attack may consist of constructing a fireline, in backburning or wetting. All of the above attributes are the same, but their values can be influenced by the type of attack. The main difference is related to the constraints that relate the attack instants and the fire arrival times.

A time horizon of T time units, the instant when the objectives are evaluated, is considered.

We define the attribute d^r that corresponds to the duration of the attack.

If there is an attack at a node, the unit can only leave the node after the duration of the attack:

$$t_p^{rub} \geq t_p^{rua} + d^r - M^r(1 - x_p^{ru}) \qquad p \in N^r \setminus \{l^r\}, r \in R, u \in U^r \tag{19}$$

We represent the set of direct attack resources as D, $D \subseteq R$, and the set of indirect attack resources as E, $E \subseteq R$, with $D \cup E = R$.

4.2.1 Direct

The constraints for direct attacks are:

$$t_p^{rua} \leq t_i^f + T(1 - x_p^{ru}) \qquad i \in N^f, r \in D, u \in U^r, p \in F_i^r \tag{20}$$

$$t_p^{rub} \geq t_i^f + d^r - T(1 - x_p^{ru}) \qquad i \in N^f, r \in D, u \in U^r, p \in F_i^r \setminus \{l^r\}. \tag{21}$$

Constraints (20) and (21) state that the unit must arrive at a node p before fire arrives at any fire node convered by it and must leave after fire is suppressed at all the covered fire nodes (the maximum instant of the fire arrival plus the duration of the attack among the covered nodes).

4.2.2 Indirect

The constraints for indirect attacks are:

$$t_p^{rua} + d^r \leq t_i^f + T(1 - x_p^{ru}) \qquad i \in N^f, r \in E, u \in U^r, p \in F_i \tag{22}$$

$$t_p^{rub} \leq t_i^f + T(1 - x_p^{ru}) \qquad i \in N^f, r \in E, u \in U^r, p \in F_i. \tag{23}$$

Constraints (22) state that the unit must arrive before fire and have enough time to implement the indirect attack (e.g., construct the fuel break). Constraints (23) state that the unit must leave before fire arrives.

5 Objectives

We consider four lexicograohic objectives: to minimize the burned *sensible* area, to minimize the burned area, to minimize the cost of the resources used, and to minimize the sum of the route durations.

The burned area is modeled with additional decision variables and constraints:

$$y_i^f = \begin{cases} 1, & \text{if node } i \text{ burns before instant } T \\ 0, & \text{otherwise} \end{cases}, \qquad i \in N^f$$

$$y_i^f \geq 1 - t_i^f / T \qquad i \in N^f.$$

Defining the set of sensible nodes as N^s, with $N^s \subseteq N^f$, the first objective is:

$$\text{Min} \sum_{i \in N^s} y_i. \tag{24}$$

The second objective is similar to (24) with the replacement of N^s by N^f.

The third objective minimizes the total cost and avoids the use of resources that are not needed. It also gives preference to attacks from a single unit than divided for more than one units.

Defining the cost of a unit of a resource r, $r \in R$, by c^r:

$$\text{Min} \sum_{r \in R, u \in U^r} c^r z^{ru}. \tag{25}$$

Lastly, the fourth objective is

$$\text{Min} \sum_{r \in R, u \in U^r} t_{lr}^{rua} - t_{lr}^{rub}. \tag{26}$$

6 Experiments

We used pyO3F, a python framework for wildfire-related optimization [12] to implement and perform some preliminary computational tests with the MIP model. These tests were conducted on a machine with a processor Intel Core i7, 1700 Mhz, 10 cores, 12 logical processors and 32 GB of physical memory. pyO3F uses fuels, winds, and slopes and the Rothermel fire behavior model to derive fire-related parameters in the landscape under study, namely the fire transmission times of the arcs fire network.

The main objective was to assess the computational difficulty in getting good quality solutions in short amounts of time in an actual landscape.

The main pyO3F parameters are given in Table 1, together with the characterization of a scenario through an ignition, an alert time, and a wind.

Figure 2 displays the landscape land use (level 1 for clarity, although level 4 was used), the fire paths and the isochronous of the fire arrival times without firefighting resources.

We considered two types of resources, one for direct attacks and one for indirect attacks, with the attributes provided in Table 2.

We conducted two experiments. In the first, we considered only one unit of the direct attack resource and a time horizon of 180 min. In the second, we added two units of the indirect attack resource and set the time horizon to 360 min.

In both cases, we set a limit for the computational time of 5 min.

Figure 3 displays the solution obtained in the first experiment. The computational time limit (5 min) was reached while optimizing the second objective. Because the third and fourth objectives are not optimized, probably the route may still be improved (not for sure because there are roads with different speeds). Since the time horizon is 180 min and the fire reaches the sensible node 'E' at 182 and the node 'NE' later, no sensible nodes burn in the time horizon: the first objective has the value 0.

Table 1 `pyO3F` scenario and main parameters

Parameter	Value	Explanation
Extent	[1468, 157830, 21563, 175535]	Extent of the area of interest
Fire resolution	2000 m	Distance between fire nodes
Roads resolution	1000 m	Distance between nodes in the roads network
Fuel source	land_use_4	How fuel information is obtained
Sensible nodes	['NE','E']	Self-explanatory
Ignition	['SE']	Self-explanatory
Alert time	60 min	Self-explanatory
Wind	('combined',60,315,0.5)	Fire rate of spread is the average between 'upslope' wind and wind in the 315o direction

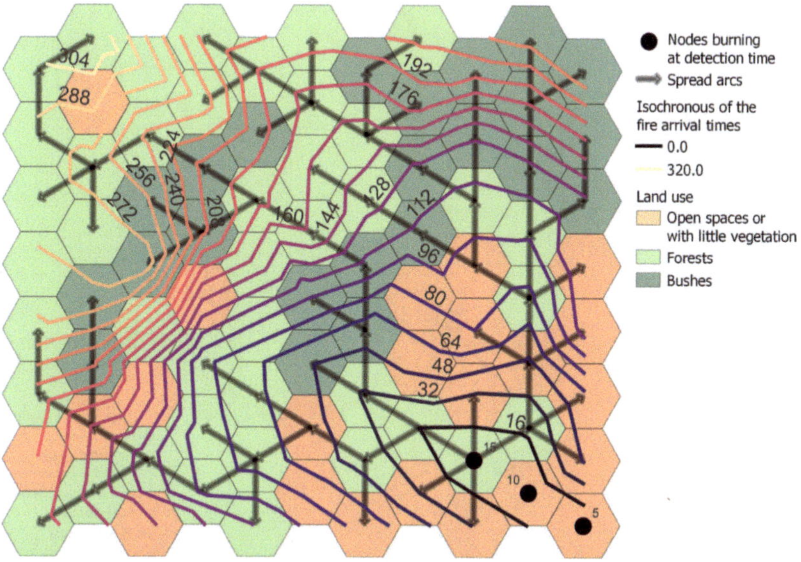

Fig. 2 Landscape representation and fire spread with no resources

Figure 4 displays the solution obtained in the second experiment. The computational time limit (5 min) was reached again while optimizing the second objective. However, it was possible to contain fire (with 50 unburned nodes out of 75) even without proving the optimality of the solution in the second objective. Fire does not reach any sensible node. Because the third objective was not optimized, it is not

Table 2 Resources characterization for direct and indirect attacks

Type of attack	direct	indirect	unit
Preparation time	30	5	min
Major roads	40	80	km/h
Minor Roads	40	60	km/h
Highway links	20	40	km/h
Very small roads	5	10	km/h
Paths unsuitable for cars	5	10	km/h
Fire delay	180	180	min
Capacity radius	5000	10000	m
Suppression radius	1000	1000	m
Suppression rate	100	50	m/min
Safety radius	500	500	m

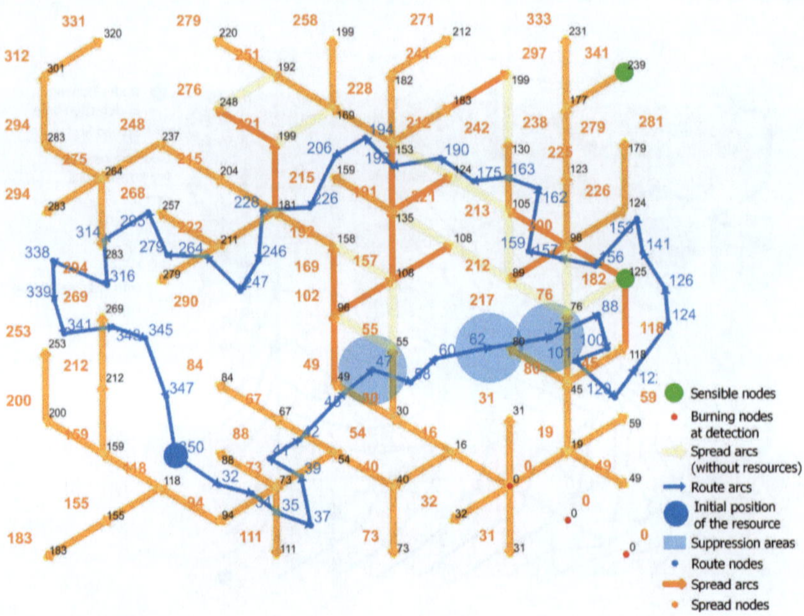

Fig. 3 Evaluation time 180 min. Solution for one direct attack resource. Numbers represent time instants of the resource arrival times (blue), fire arrival times without resources (in black) and with resources (in orange). Computational time limit (5 min) was reached, while optimizing the second objective

clear if fewer units could be used. Because the fourth objective was not optimized, the routes could be improved, at least by removing the attacks where the fire is contained.

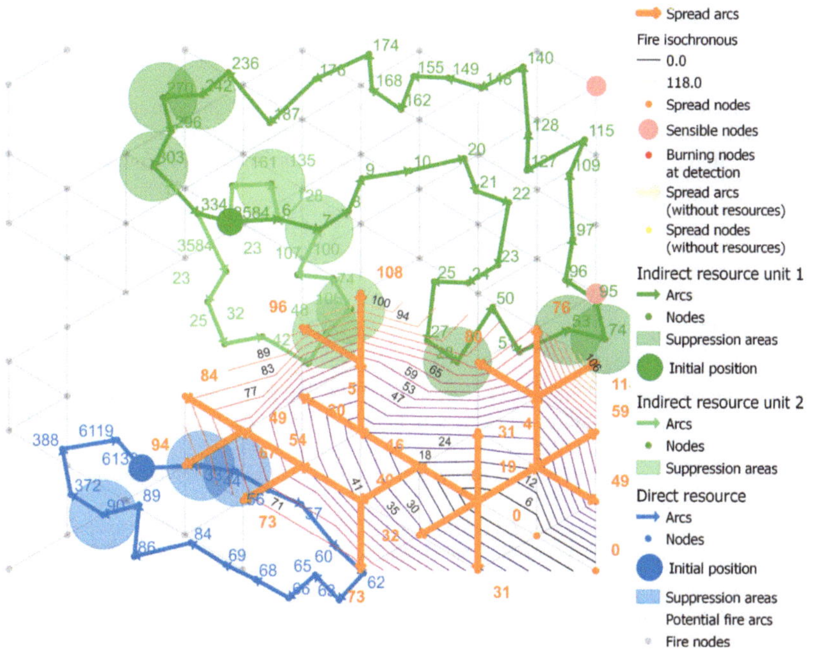

Fig. 4 Evaluation time 360 min. One direct and two indirect resources available. Computational time limit (5 min) was reached, while optimizing the second objective. Fire is contained because all the fire paths end at attacked nodes or at nodes without burnable outgoing arcs

7 Conclusions

In this paper, we described an MIP model for defining the routes and attack positions of firefighting resources to mitigate the impact of a wildfire. The MIP model is based on decision variables that characterize the fire spread and others that characterize the route and attack positions of the units of resources. Time constraints relating fire and resources ensure that firefighting resources follow safe paths.

Preliminary experiments, with an implementation of the model in the `pyO3F` framework, allowed to validate the implementation and to assess the quality of the solutions in a actual landscape represented by a fire network with 75 nodes and 330 arcs and a road network for potential movements and attacks positions of 252 nodes and 1038 arcs. For a small computational time (5 min), it was possible to obtain solutions with attack positions with an effective impact on protecting sensible nodes and minimize the number of burned nodes, although with (what it seems to be) long routes and unnecessary attack positions.

This work may be used as the basis for an iterative approach in steps, similar to the MIP heuristic proposed in [6], where an initial attack is modelled by limiting to one the number of attacks of each unit of each resource. Additionally, given the almost real-time requirements to solve the problem in practice, (meta-) heuristics should also be explored in the future.

Acknowledgements This work has been supported by FCT—Fundação para a Ciência e Tecnologia within the R&D Unit Project Scope UID/00319/Centro ALGORITMI (ALGORITMI/UM) and within project PCIF/GRF/0141/2019 "O3F—An Optimization Framework to Reduce Forest Fire".

References

1. Cunningham, C.X., Williamson, G.J., Bowman, D.M.J.S.: Increasing frequency and intensity of the most extreme wildfires on earth. Nat. Ecol. Evol. **8**, 1420–1425 (2024). https://doi.org/10.1038/s41559-024-02452-2
2. Martell, D.L.: A review of operational research studies in forest fire management. Can. J. For. Res. **12**(2), 119–140 (1982)
3. Hof, J., Omi, P.N., Bevers, M., Laven, R.D.: A timing-oriented approach to spatial allocation of fire management effort. For. Sci. **46**(3), 442–451 (2000). https://doi.org/10.1093/forestscience/46.3.442
4. Finney, M.A.: Fire growth using minimum travel time methods. Can. J. For. Res. **32**(8), 1420–1424 (2002). https://doi.org/10.1139/x02-068
5. Alvelos, F.: Mixed integer programming models for fire fighting. In: International Conference on Computational Science and Its Applications, pp. 637–652. Springer International Publishing, Cham (2018). https://doi.org/10.1007/978-3-319-95165-2_45
6. Alvelos, F., Marto, M., Mendes, A.: Dispatching, positioning and routing resources for wildfire initial attack. Netw. **85**(1), 80–102 (2025). https://doi.org/10.1002/net.22278
7. Wei, Y., Rideout, D.B., Hall, T.B.: Toward efficient management of large fires: a mixed integer programming model and two iterative approaches. For. Sci. **57**(5), 435–447 (2011). https://doi.org/10.1093/forestscience/57.5.435
8. Belval, E.J., Wei, Y., Bevers, M.: A mixed integer program to model spatial wildfire behavior and suppression placement decisions. Can. J. For. Res. **45**(4), 384–393 (2015)
9. Mendes, A.B., Alvelos, F.: Iterated local search for the placement of wildland fire suppression resources. Eur. J. Oper. Res. **304**(3), 887–900 (2023). https://doi.org/10.1016/j.ejor.2022.04.037
10. Granda, B., Vitoriano, B. (2024). The team orienteering problem with variable time windows. International Transactions in Operational Research
11. Granda, B., León, J., Vitoriano, B., Hearne, J.: Decision support models and methodologies for fire suppression. Fire **6**(2), 37 (2023)
12. Alvelos F.: pyO3F—A python framework for wildfire-related optimization, Part I: Design and fundamentals. In: Alvelos, F., Martins, I., Rocha, A. (eds.) Advances in Optimization and Wildfire: Proceedings of the First Optimization and Wildfire Conference, October 1–4, 2024, Luso, Portugal. Springer (2025).

Integrating Fire Suppression in Forest Management

Filipe Alvelos and Isabel Martins

Abstract This work proposes a more holistic forest management approach compared to previous efforts by the same authors, by integrating suppression planning into an already developed fire simulation-based optimization framework. Forest management is iteratively guided by feedback from fire spread simulations (developed in previous work) and the optimization of firefighting resource base locations. The approach employs mixed integer programming to solve the forest management problem and optimize the distribution of firefighting bases. It uses shortest path algorithms to assess fire potential by identifying high-risk paths that could lead to unacceptable fire spread. If such paths are detected, the algorithm adjusts the constraints of the forestry model and re-optimizes both forest management and firefighting base placement. This iterative process continues until no unacceptable fire paths are found. Fire paths are obtained by representing the forest as a network. To model potential base locations, a resource network is also defined. We present computational experiments conducted in a Portuguese forest to test this approach.

1 Introduction

Just as fire plays a fundamental role in forest ecosystems, managing fires is an essential component of overall forest management. The different tasks that comprise fire management can be roughly grouped into the following categories: fuel management, fire prevention, detection, suppression resource acquisition, deployment, dispatch and use, and large fire management. Fuel management and suppression planning are

F. Alvelos (✉)
Department of Production and Systems/ALGORITMI Research Centre/LASI, School of Engineering, University of Minho, Braga, Portugal
e-mail: falvelos@dps.uminho.pt

I. Martins
School of Agriculture, University of Lisbon/Center for Mathematical Studies, University of Lisbon (CEMS.UL), Lisbon, Portugal
e-mail: isabelinha@isa.ulisboa.pt

F. Alvelos et al. (eds.), *Advances in Optimization and Wildfire*, Lecture Notes in Operations Research, https://doi.org/10.1007/978-3-032-03108-2_6

closely interconnected aspects of fire management, as land use significantly influences fuel profiles and subsequent wildfire behaviour [1]. Fuel management is the planned manipulation or reduction of living or dead forest fuels for forest management and other land use objectives (e.g. hazard reduction, silvicultural purposes, wildlife habitat improvement) by prescribed fire, mechanical, chemical, or biological means, or by changing stand structure and species composition. Suppression involves extinguishing a wildfire, preventing or modifying the movement of an unwanted fire, or managing a fire when it provides benefits like vegetation reduction or improved wildlife habitat. Firefighters control the spread of a fire primarily by removing its sources of heat and fuel. They remove heat by applying water or fire retardant on the ground with pumps or wildland fire engines, or by air using helicopters or airplanes. They reduce fuel by removing burnable vegetation with hand tools, using heavy equipment like bulldozers, and setting controlled fires. Fire suppression efforts can be categorized into two primary subsystems: initial attack and extended attack, also known as large fire management. The initial attack involves the early stage of suppression efforts, during which fire agencies aim to control fires while they remain small. If a fire's size and intensity exceed the capacity of initial attack resources, it is classified as an escaped fire. Despite the interdependence between fuel management and suppression planning, previous optimization models for forest fire management have often treated these components separately [2].

In Portugal, timber production is the main contributor to the gross value of the forest sector, positioning it as a core focus of forest management. The main goal is often to optimize the net present value of forestry operations over time through practices such as clearcutting, thinning, planting, fertilization, and shrub clearing. Consequently, fuel management becomes an integral component of these forest management practices. In this study, we propose explicitly integrating forest management for timber production and fuel management activities with suppression decisions. To the best of our knowledge, this is the first work to address these aspects in a unified framework.

This paper is summarized as follows. In Sect. 2, we present a literature review. In Sect. 3, we define the problem and outline the main aspects of the modelling. In Sect. 4, we provide details on the proposed approach. In Sect. 5, we report on computational experience regarding the efficiency of the approach. A case study from an actual landscape is used to illustrate the proposed method and derive insights. In the last section, we present our conclusions.

2 Related Work

This review explores the development of studies that aim to integrate fuel management with suppression decisions using operations research techniques. The list of references up to 2015 is based on the work of [2], who identified two prior contributions in this area: [3], whose approach was not fully integrated, and [4], whose model lacked explicit consideration of suppression decisions. Furthermore, we did not identify more recent studies that integrate these components, nor did the review by [6].

In [3], the authors developed a mixed integer linear programming model to evaluate trade-offs in various fuel treatment scenarios. The model optimizes two objective functions: the cost of deploying initial-attack resources and the expected cost of managing fires that escape initial attack. By minimizing a weighted sum of these objectives, it quantifies the trade-off between investments in fuel treatments, initial attack resources, and the costs associated with containing escaped fires. The model relies on data from fire scenarios, each representing a set of fire locations occurring within a single day. Each fire location is characterized by parameters that quantify the reduction in escape risk achieved by increasing the number of resources dispatched for initial attack. These escape risk reductions depend on the fuel treatment scenario. The model includes integer decision variables for the number of suppression resources assigned to each station and the number of resources dispatched from each station to fires. The model also incorporates the probability of fire escape at each location and the probability of each fire scenario occurring, ensuring that the analysis accounts for uncertainty in fire behaviour and occurrence. The model was applied to a hypothetical problem involving a 10×10 grid of forest districts that covers 2497 hectares.

In [4], the author proposed a mixed integer programming formulation to select fuel treatment locations within a planning period. The model simultaneously accounts for fire ignitions from multiple potential locations throughout a landscape, with the aim of minimizing total future fire loss. The objective function incorporates the loss caused by each fire ignition, weighted by the probability of the ignition occurring and the probability of it persisting for a given duration. Fuel treatments are selected with the expectation that they will enhance opportunities for successful fire control in the event of future fires. However, the model does not consider dispatch issues.

The work [2] describes an integer programming model that integrates fuel management and suppression preparedness decisions, offering an integrated planning framework. In this study, the landscape is divided into a series of cells that represent potential fire locations and candidate fuel treatment sites. These cells can vary in shape and size. Additionally, a set of potential bases for deploying suppression resources is defined, which may include existing permanent and temporary bases, as well as locations suitable for "forward deployment" of suppression resources. Primary decisions involve determining where to base suppression resources and where to implement fuel treatments. The objective function maximizes the weighted number of cells covered. Cell weights can be assigned based on factors such as ignition probability and the values at risk if a fire originating in that cell is not contained by initial attack resources. Although this integrated optimization model represents a significant advancement in combining fuel management and fire suppression preparedness planning, it was only tested on hypothetical 100-cell landscapes, limiting insights into its practical performance in real-world scenarios.

Before concluding this section, it is worth noting that the present work integrates suppression planning using the fire simulation-based optimization framework developed in [5].

3 Problem Definition and Modelling Main Aspects

3.1 Problem Overview

We consider a forest divided into stands, that is, geographically continuous parcels of land sufficiently uniform in their characteristics (e.g., species composition) to be distinguished from adjacent areas. For each stand, a set of prescriptions, i.e. a set of silviculture actions (e.g. thinning and harvesting) to be taken over a time horizon, is considered. An example of a prescription for a stand with Eucalyptus never harvested is to perform a first harvest at 12 years, followed by harvests every 10 years. We also consider different alternatives for the periodicity of shrub cleaning. A combination of a prescription and a shrub cleaning results in a *plan*. Each plan is characterized by annual estimates of the Net Present Value (NPV) resulting from forestry operations, as well as estimates of three ecosystem services, carbon stock, biodiversity, and soil erosion.

The problem consists in selecting one plan for each stand to maximize NPV taking into account annual bounds on:

- ecosystem services;
- fire potential.

Furthermore, a minimum standing timber volume is required at the end of the planning horizon to prevent overharvesting, along with lower and upper bounds on annual harvest volumes to maintain supply.

We also consider that firefighting resources are available and that their base(s) (i. e., their location when an ignition occurs) can be different each year. The existence of these firefighting resources allows one to reduce the fire potential near the base(s) as it is assumed that, if the ignition is within a given time from the base, the initial attack is successful.

In the next two subsections, we describe how fire potential and fire suppression are modelled.

3.2 Fire Potential Estimate

The measure of fire potential used is the maximum fire rate of spread (ROS) among a representative subset of fire paths originating from a large set of independent ignitions. Consequently, in addition to meeting the constraints on ecosystem services and timber volumes, a solution (that is, the allocation of a specific management plan for each stand) is considered feasible only if no fire paths exhibit a ROS exceeding a specified threshold (ROS_{max}).

Fire paths are obtained by representing the forest as a network and using the minimum travel time (MTT) principle [7]. The nodes correspond to cells, and the arcs correspond to potential fire transmissions between adjacent cells. All cells are assumed to be of the same size, and a forest stand may comprise multiple cells. Each arc is associated with an estimate of the fire transmission time between its start and end nodes. The MTT principle states that the fire path between nodes is the quickest path with respect to fire transmission times. Given one ignition node, fire spread (that is, fire arrival times at all nodes and the fire path between the ignition node and all other nodes) can be obtained by a shortest path tree algorithm (e.g. Dijkstra's algorithm).

The management plans of a solution influence the fire transmission times and, therefore, the fire paths. More precisely, each year of each plan is associated with a fuel model (i.e., a set of parameters that characterizes the fuel used as input for fire behaviour models).

The fire potential of a solution is obtained by simulating the fire spread in all years and all possible ignitions. In the proposed approach, the fire transmission times are obtained using the computational implementation of the Rothermel model [8] of Behave Plus 6 software [9]. The inputs for the model include the upslope wind (wind aligned with slope), the slope, and the fuel models specifically developed for Portugal [10, 11].

3.3 Integrating Fire Suppression

The presence of firefighting resources allows one to reduce fire potential. We assume that, each year, a given number of bases for initial attack terrestrial firefighting resources can be placed at the forest. For modelling the potential locations for the bases and the movement of the resources, we define a resource network. Different firefighting resources can be considered simultaneously. They can be aerial or terrestrial and, among each, they can be distinguished by their speed in each arc and the coverage radius.

If the quickest path from the base to a resource node takes less than a given parameter, the resource node is termed reachable. Firefighting suppression occurs in fire nodes in an area centred on the resource node where the firefighting resource is placed. Those fire nodes are said to be covered by that resource node. A fire node that is covered from a resource node that is reachable from a base is termed *attainable* from that base. If a fire node is attainable by at least one base, it is termed *(fire) barrier node.* Fire does not spread from barrier nodes and therefore fire paths do not include barrier nodes. The appropriate location of the bases generally reduces the fire potential. Figure 1 illustrates these concepts.

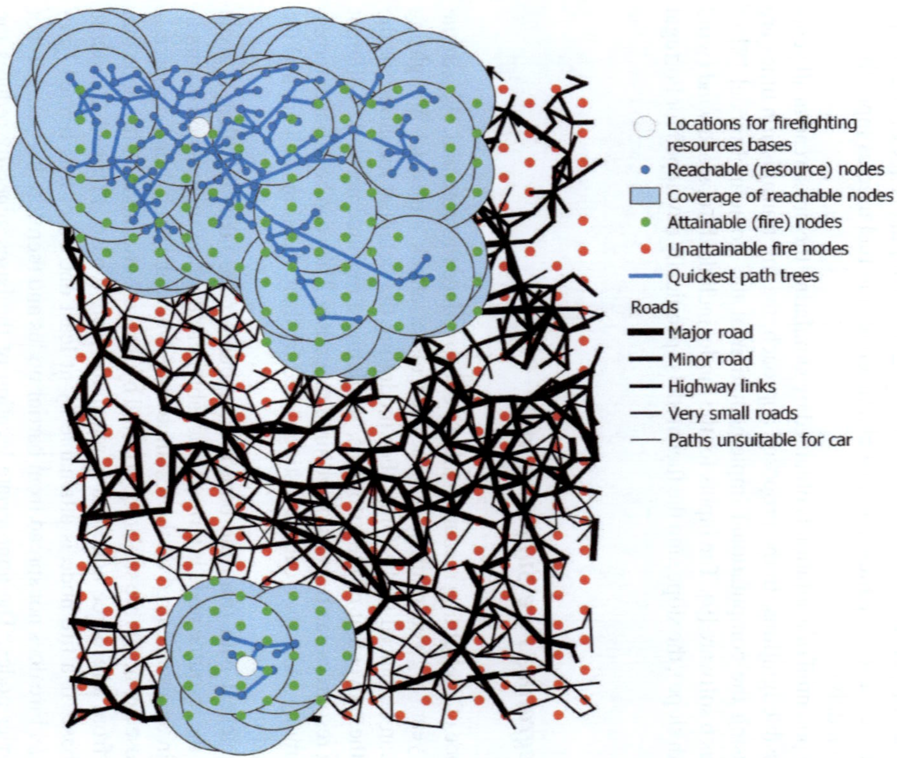

Fig. 1 Illustration of how fire barrier (attainable) nodes are defined with two bases for terrestrial resources. The fire nodes correspond to the centers of hexagons that constitute a tiling

4 Proposed Approach

4.1 Overview

As detailed in the previous section, the problem involves allocating a management plan to each stand and determining the annual locations of a given number of bases. This must be done while respecting annual constraints on ecosystem services, timber volumes, and fire potential, with the objective of maximizing NPV. The proposed approach to address this problem has a main cycle comprising five steps. Algorithm 1 provides an overview of the method. The stop criterion is checked on lines 5 and 10. We detail each main step in the following subsections.

Algorithm 1 Algorithmic overview of the proposed method.

```
     iteration = 0
     repeat
         Solve forestry problem with MIP
         Fire potential assessment through shortest path algorithms
 5:      if No unacceptable paths then
             Break
         end if
         For each year, solve the location of firefighting suppression resources with MIP
         Fire potential reassessment through shortest path algorithms
10:      if No unacceptable paths then
             Break
         end if
         Update forestry problem with constraints corresponding to unacceptable paths
         iteration = iteration + 1
15: until Time limit is reached
```

4.2 Forestry Problem

In the first iteration, the first step involves solving a forestry problem using a mixed integer programming model (MIP) to obtain an optimal solution (that is, an allocation of management plans to stands) without considering fire risk. This solution maximizes NPV taking into account the ecosystem service constraints (that is, annual lower bounds on the biodiversity indicator and carbon stock, and annual upper bounds on soil erosion). It also respects the constraints on the final standing timber volume and the annual harvested timber volume, but does not consider the fire potential threshold. In the remaining iterations, fire-related constraints will be added to the MIP as described below.

Details of the forestry problem can be found in [5]. In the same reference, other aspects are addressed, such as the use of stands that remove all fire paths where they are included (stands that do not contribute to the NPV or services, referred to as *breaking stands*) and mixed stands, where less than 75% of the trees in the canopy are of single species. Although these aspects are also incorporated into this approach, given that the focus of this paper is on the integration of fire suppression resources, we refer interested readers to that reference.

4.3 Fire Potential Assessment

In the second step, fire potential is assessed by identifying unacceptable paths – those with a ROS higher than the given threshold – using shortest path algorithms. A fire network is constructed for each year, with fire transmission times derived from the fuel model of each node based on the management plan of the stand it belongs to. Wind and slope, which remain constant throughout the execution of the method, are also taken into account as previously described. If there are no acceptable paths, the method stops with the current solution. The identification of unacceptable paths is based on the assumption that an ignition can occur at any node of the fire network. We limit the search for unacceptable paths to the set P, the set of fire paths longer than the threshold value d and that are subpaths of other paths, in order to efficiently reduce the search space of the firefighting resource location problem and focus only on those paths most likely to contribute to constraint violations. Nevertheless, a high value of d can hide unacceptable subpaths with high ROS and a considerable length. A small value of d may identify as unacceptable, paths with a high ROS but only in a small part of their length. The ROS of path p is calculated as the ratio of its length to the total fire transmission time across all its arcs. The set P can be obtained by enumeration or by applying shortest path algorithms (e.g. repeating Dijkstra by changing the source or Floyd–Warshall).

4.4 Locating Firefighting Suppression Resources

If unacceptable paths are identified in the fire potential assessment step, the firefighting suppression resources base location problem is solved. This problem consists of selecting the location of a fixed number of bases, where the initial attack firefighting resources are placed each year, to minimize the maximum ROS of the unbroken unacceptable paths, that is, the unacceptable paths that are not broken by an initial attack of the firefighting resources of a base. An unacceptable path is *broken* if one of its fire nodes is a barrier node (as defined in Sect. 3.3). Given the expected existence of alternative optimal solutions, a second objective (in the lexicographic sense) is considered: to maximize the number of fire barrier nodes. The lexicographic MIP model with two objectives is now presented.

Let K the set of potential locations for bases and b the number of available bases. Let P^u be the set of unacceptable paths before the bases are located and N^u the set of nodes belonging to at least one path in P^u. The ROS of unacceptable path $p, p \in P^u$, is represented by r_p. Let S_i be the subset of potential locations for bases ($S_i \subseteq K$) such that the fire node i, $i \in N^u$, is a barrier node.

The decision variables of the model are as follows.

$$y_k = \begin{cases} 1, & \text{if a suppression base is located at } k \\ 0, & \text{otherwise} \end{cases} ; k \in K$$

$$u_p = \begin{cases} 1, & \text{if at least one node of the unacceptable path } p \text{ is a barrier node} \\ 0, & \text{otherwise} \end{cases} ; p \in P^u$$

$$v_i = \begin{cases} 1, & \text{if node } i \text{ is barrier node} \\ 0, & \text{otherwise} \end{cases} ; i \in N^u$$

z – maximum ROS of an unbroken unacceptable path.

The model is as follows.

$$Lex$$

$$Min\, z \tag{1}$$

$$Max \sum_{i \in N^u} v_i \tag{2}$$

subject to

$$\sum_{k \in K} y_k \le b \tag{3}$$

$$v_i \le \sum_{k \in S_i} y_k \qquad i \in N^u \tag{4}$$

$$u_p \le \sum_{i \in p} v_i \qquad p \in P^u \tag{5}$$

$$z \ge r_p(1 - u_p) \qquad p \in P^u \tag{6}$$

$$y_k \in \{0, 1\}, k \in K; u_p \in \{0, 1\}; p \in P^u; v_i \in \{0, 1\}; i \in N^u; z \ge 0. \tag{7}$$

The explanation of this model is the following.

- (1)—Minimize the maximum ROS among all unbroken unacceptable paths.
- (2)—Maximize the number of fire barrier nodes.
- (3)—The number of locations for bases is limited.
- (4)—A node can be a barrier node only if it is attainable at least from a base location.

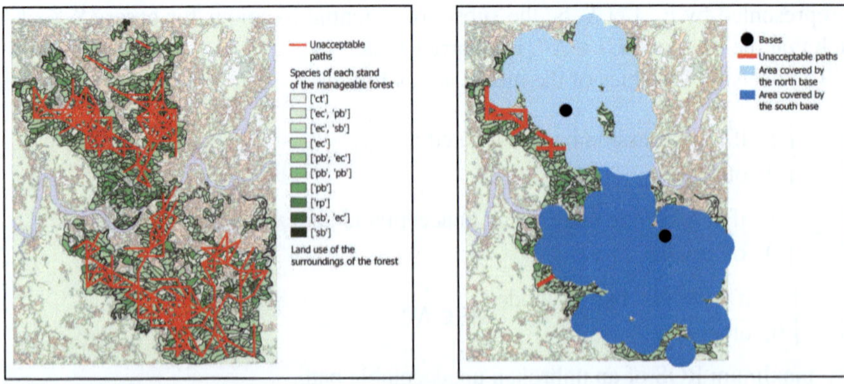

(a) Before locating bases. (b) After locating two bases optimally.

Fig. 2 Unacceptable paths for a given year and ROS threshold

- (5)—These constraints establish the relationship between the variables u_p and v_i. For each unacceptable path p, if some node in p is a barrier node (i.e., $\sum_{i \in p} v_i \geq 1$), the model enforces $u_p = 1$ in an optimal solution due to constraints (6), the binary nature of u_p and the objective function (1) that minimizes z. If there are no barrier nodes in p, then $u_p = 0$.
- (6)—These constraints, taking also into account the objective function, state that the maximum ROS is the ROS of an unbroken unacceptable path and has the largest value among such paths.
- (7)—The domain of the variables is defined.

Figure 2 shows the set of unacceptable paths before and after solving the location of suppression resources problem.

4.5 Fire Potential Reassessment

The fire potential needs to be reassessed because the fire finds other paths to spread avoiding the barrier nodes. Some unacceptable paths are broken by the bases, but some others may appear. Therefore, the shortest path algorithms for identifying unacceptable paths are run again but in a network where all outgoing arcs from an attainable node are removed. If no unacceptable paths are identified, the method stops with the current solution.

4.6 *Forestry Problem Update*

The implication for the forestry problem of an unacceptable path is that the set of stands associated with that fire path cannot simultaneously follow the currently selected management plans. Thus, we introduce a constraint to ensure that not all of these stands can retain their selected plan (or, more effectively, their selected plan or any alternative plan with the same fuel model). Details of this reasoning and a formal description are provided in [5].

5 Experiments

Computations were performed on a desktop computer with an 11th Gen Intel(R) Core(TM) i9-11900 2.50 GHz processor and with 32.0 GB RAM. In Gurobi, we used the default value of the MIPgap parameter and set a time limit of 24 h for the overall method and one hour for each optimization. Thus, the solver ends with an optimal result for each problem whenever the gap between the lower and upper objective bounds is less than 0.01% times the absolute value of the incumbent objective value, or when the time limit is reached.

5.1 *Overview*

The study focuses on a forest area located in the north of Portugal, which includes two *Zonas de Intervenção Florestal* (Forest Intervention Areas), ZIF Entre Douro e Sousa, ZIF-EDS, and ZIF Paiva, ZIF-Pv, spanning 14313 hectares and dominated by eucalyptus in pure and mixed stands, along with other species like chestnut, cork oak, and riparian zones. Over a 50-year planning horizon, 85572 forest management plans were simulated for 1406 stands, involving species conversion and fuel treatment scenarios (no shrub cleaning, 5 or 10 years for shrub cleaning periodicity), along with assessments of ecosystem services like soil erosion, biodiversity, and carbon storage. Plans are based on species-specific growth models, ensuring plantations after species rotations, or natural regeneration for cork oak, with net present values calculated at a 3% discount rate. Fuel models were assigned to predict fire behaviour based on vegetation characteristics. See [5] for a more detailed description of the case study.

As for the resources, we only considered one terrestrial resource type with speeds of 40, 40, 20, 5 and 5 km/h on five types of roads / paths from major roads to paths unsuitable for cars. A fire node is attainable if it is within a radius of 1 km of a resource node reachable in 15 min from a base. The parameter d used to identify unacceptable paths was set at 2 km.

We carried out a series of experiments to evaluate the effectiveness of the proposed approach in supporting forest management. We used the same parameter values for the forest model as described in [5] and used the framework for implementing fire-related optimization approaches pyO3F [12, 13] with a resolution of 1 km for both the fire and the resource network. The extent has 17.4 km by 22.6 km, and the number of nodes of the fire network is 377.

We performed three primary experiments, *A*, *B*, and *C*, where the number of bases available is set at 0, 1 and 2, respectively. Each experiment incorporated the fuel models of the surrounding areas, as well as constraints related to biodiversity, erosion, carbon stock, final standing timber volume, and annual harvested timber volume. In this context, we present the impact on the NPV, ecosystem services, the total volume of timber removed, the final standing volume, as well as an overview of species composition at the end of the planning horizon and the periodicity of shrub clearing. In all cases, the fire was introduced by gradually reducing ROS_{max} (the threshold for ROS), representing a forest with progressively increasing fire resilience. Refer to [5] for details on the parameters related to the integration of forest management and fire simulation.

5.2 Results and Discussion

Table 1 shows an optimal solution of the forest management problem for each experiment. Figure 3 illustrates the relationship between total NPV, NPV-T, and ROS_{max} for experiments *A*, *B* and *C*, which correspond to different numbers of available bases (0, 1, and 2, respectively). This relationship can support decision-making regarding whether to increase the number of bases, depending on the level of risk the decision-maker is willing to assume (i.e., the maximum acceptable ROS) and the potential gain in NPV. For example, at a $ROS_{max} = 6$ km/h, having zero or one base results in a difference of approximately 9000 € in NPV.

For $ROS_{max} = 8$ km/h, we observed that, for all experiments, there were no unacceptable fire paths and the largest ROS in the forest concerning the optimal solution provided by the forestry model was 7.5 km/h. We aso observed that all solutions obtained for ROS_{max} equal to 8 and 7 Km/h in all experiments, as well as for $ROS_{max} \leq 4$ Km/h in experiment *A*, are optimal under Gurobi's default relative gap tolerance. The remaining solutions were obtained with relative gaps ranging from 0.0104% to 0.0261%.

For each experiment j, $j = A, B, C$, we considered two relative variations of the NPV according to ROS_{max} (Table 2): the gap between the NPV obtained for the current ROS_{max} and the NPV for ROS_{max} set at 8 km/h, 1RNPV-T^j,

$$1RNPV - T^j = \frac{NPV - T^j forROS_{max} = i - NPV - T^j forROS_{max} = 8}{NPV - T^j forROS_{max} = i}, \text{ for } i = 7, .., 1$$

and the gap between the NPV obtained for the current ROS_{max} and the NPV for the previous ROS_{max}, 2RNPV-T^j,

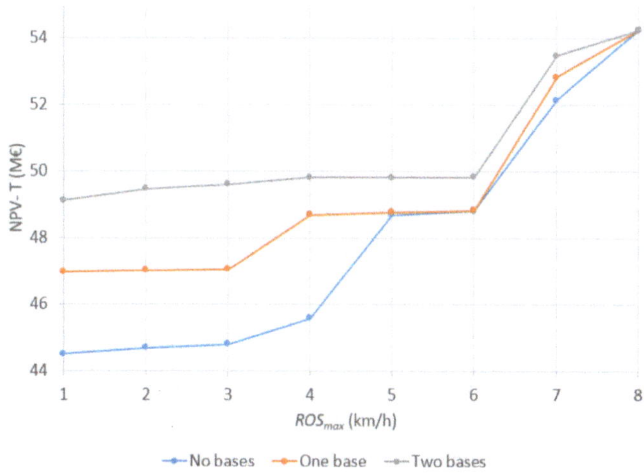

Fig. 3 NPV-T (total NPV) vs. ROS_{max} for zero, one, and two bases

$$2RNPV - T^j = \frac{NPV - T^j forROS_{max} = i - NPV - T^j forROS_{max} = i + 1}{NPV - T^j forROS_{max} = i}, \text{ for } i = 7, .., 1.$$

With 1RNPV-T^j, we can determine the variation, in percentage, of the NPV compared to a scenario without considerations of fire (e.g. where $ROS_{max} = 8$ km/h). In contrast, for 2RNPV-T^j, the variation is calculated relative to a scenario in which the forest is sensitive to more severe fires, specifically, those with a ROS less than the current value ROS_{max} plus 1 km/h.

For each experiment $j, j = B, C$, we also defined 3RNPV-T^j (Table 2),

$$3RNPV - T^j = \frac{NPV - T^j forROS_{max} = i - NPV - T^A forROS_{max} = i}{NPV - T^j forROS_{max} = i}, \text{ for } i = 8, .., 1.$$

Using 3RNPV-T^j, we also considered the relative impact of addressing firefighting suppression resources bases (experiments B and C) compared to a no-suppression situation (experiment A).

Tables 3 and 4 complete the information: Table 3 characterizes the stands in terms of species in year 50 (for each pair $X - Y$, species X is dominant (in mixed stands)); Table 4 displays the stand area assigned to each shrub cleaning periodicity (for each pair $Z - W$, periodicity Z is relative to the dominant species (in mixed stands)).

Integrating suppression generally reduces the number of unacceptable fire paths (as indicated by the number of cuts in Table 1), thereby increasing the availability of management plans that are more attractive in terms of NPV (the objective maximized in the forestry management problem). The consequences of this are reflected in Tables 1–4.

Table 1. Integrating suppression into forestry management improves both the NPV and the volume of harvested timber, while reducing the number of breaking

Table 1 Optimal solutions of the forestry problem concerning all experiments. NPV-T—total NPV; $\bar{b}$—average biodiversity by year; $\bar{e}$—average erosion by year; $\bar{c}$—average stock of carbon by year; Vol-T—total volume of removed timber; Stand. vol.—standing timber volume; # iter.—number of iterations; # cuts—number of added unacceptable path constraints; $ForP_{curr}$ - forestry problem

ROS_{max} (C)	NPV-T (€)	$\bar{b}$ (/ha)	$\bar{e}$ (t/ha)	$\bar{c}$ (t/ha)	# breaking stands	Vol-T (m^3)	Stand. vol. in T (m^3)	# iter.	# cuts
Experiment *A* (elapsed time—52462 s; time to optimize $ForP_{curr}$ 51450 s; time to determine unaceptable paths 167 s; time to convert unacceptable paths to MIP constraints 3 s)									
8	54246880	2.090	8.894	23.176	0	6274086	669802	1	0
7	52143107	2.002	8.614	22.248	48	6033416	660194	29	+1159
6	48812456	1.976	8.474	21.602	69	5734350	593880	4	+650
5	48682065	1.950	8.362	21.302	78	5688403	597366	1	+136
4	45577028	1.864	8.092	20.514	121	5298877	581620	17	+506
3	44788085	1.858	8.024	20.224	129	5204184	584666	17	+32
2	44680116	1.854	8.030	20.160	130	5195771	586214	10	+13
1	44505553	1.852	8.020	20.022	130	5182991	583215	5	+5
Experiment *B* (elapsed time—51480 s; time to optimize $ForP_{curr}$ 50384 s; time to determine unaceptable paths 76 s; time to convert unacceptable paths to MIP constraints 1 s)									
8	54246880	2.090	8.894	23.176	0	6274086	669802	1	0
7	52843754	2.034	8.708	22.582	31	6111595	691622	29	+772
6	48821551	1.964	8.432	21.326	69	5746438	601592	8	+842
5	48775499	1.964	8.442	21.432	71	5739396	602524	1	+45
4	48679597	1.966	8.434	21.396	73	5715086	617235	1	+61
3	47039817	1.892	8.274	20.614	99	5520155	576521	10	+205
2	47015471	1.890	8.270	20.670	100	5517180	575028	4	+12
1	46974515	1.890	8.248	20.656	102	5502325	576556	5	+14
Experiment *C* (elapsed time—34200 s; time to optimize $ForP_{curr}$ 33279 s; time to determine unaceptable paths 55 s; time to convert unacceptable paths to MIP constraints 1 s)									
8	54246880	2.090	8.894	23.176	0	6274086	669802	1	0
7	53475954	2.050	8.750	22.798	0	6177353	677615	25	+482
6	49820637	1.984	8.474	21.680	61	5831317	617214	7	+779
5	49814040	1.974	8.462	21.512	63	5817299	622164	1	+47
4	49806793	1.980	8.436	21.642	65	5804585	625448	1	+58
3	49599626	1.970	8.384	21.702	69	5792781	605827	1	+70
2	49458254	1.948	8.368	21.476	74	5765631	606859	1	+53
1	49116143	1.928	8.302	21.170	85	5707549	597594	1	+24

Table 2 Relative variations of the optimal NPV, 1RNPV-T^j and 2RNPV-T^j, for $j = A, B, C$, and 3RNPV-T^j, for $j = B, C$

	1RNPV-T^j			2RNPV-T^j			3RNPV-T^j	
ROS_{max}	A	B	C	A	B	C	B	C
(km/h)	(%)	(%)	(%)	(%)	(%)	(%)	(%)	(%)
8	–	–	–	–	–	–	0.00	0.00
7	−4.04	−2.66	−1.44	−4.04	−2.66	−1.44	1.33	2.49
6	−11.13	−11.11	−8.89	−6.82	−8.24	−7.34	0.02	2.02
5	−11.43	−11.22	−8.90	−0.27	−0.09	−0.01	0.19	2.27
4	−19.02	−11.44	−8.92	−6.81	−0.20	−0.02	6.37	8.49
3	−21.12	−15.32	−9.37	−1.76	−3.49	−0.42	4.79	9.70
2	−21.41	−15.38	−9.68	−0.24	−0.05	−0.29	4.97	9.66
1	−21.89	−15.48	−10.45	−0.39	−0.09	−0.70	5.26	9.39

stands. These effects become more pronounced when the number of suppression bases increases to two. Biodiversity, erosion, and carbon storage indicators also improve with the addition of bases across nearly all ROS_{max} values. The volume of standing timber at the end of the planning horizon increases with suppression integration up to $ROS_{max} = 4$ km/h in the single-base scenario, but then declines compared to the experiment without suppression. In contrast, with two bases, the standing volume consistently increases across all ROS_{max} values when compared to the other experiments. The computational time required for solving the forestry optimization problem decreases as the number of suppression bases increases.

Table 2. 1RNPV-T^j: for each fire resilience level (ROS_{max}), the relative decrease in NPV—when compared to a scenario without fire considerations ($ROS_{max} = 8$ km/h)—is less pronounced with one suppression base, and even smaller with two bases, than with no suppression at all. No clear pattern emerges for the 2RNPV-T^j. 3RNPV-T^j: however, we observe a consistent relative increase in NPV when moving from no bases to one base, and from no bases to two bases; the most substantial improvement occurs in the latter case, particularly for $ROS_{max} \leq 4$ km/h.

Table 3. The inclusion of suppression, combined with an increase in the number of bases, led to an expansion of eucalyptus and pine areas by the end of the planning horizon, while cork oak areas declined. Chestnut area increased when the number of bases was set to two but decreased when only one base was considered, for certain values of ROS_{max}—compared to the experiment without suppression integration. Greater availability of management plans that are more attractive in terms of NPV typically favors species such as eucalyptus and pine.

Table 4. Integrating suppression led to an increase in areas without shrub clearing, especially for $ROS_{max} \leq 4$ km/h. This relates to the previous discussion: management plans that are more attractive in terms of NPV are those that avoid shrub clearing costs.

Table 3 Stand composition in terms of species, concerning the last year of the planning horizon, for all experiments. Ct—chestnut; Ec—eucalyptus; Pb—maritime pine; Sb—cork oak; Qr—pedunculate oak; Rp—riparian

ROS_{max}	Year 50, Area (ha)													
(km/h)	Ct-Ec	Ct	Ec-Pb	Ec-Sb	Ec	Pb-Ec	Pb-Sb	Pb	Qr	Rp	Sb-Ec	Sb-Pb	Sb	Total
Experiment *A*														
8	11.70	261.46	691.49	55.69	9579.55	275.37	17.95	2186.80	0	111.62	0	9.61	1112.69	14313.91
7	11.71	250.08	680.46	55.69	9180.44	280.81	12.50	2065.55	0	111.62	0	9.61	1064.13	13722.59
6	11.71	247.94	631.47	55.69	8409.11	261.45	3.17	2348.35	0	111.62	20.71	13.98	1132.60	13247.75
5	11.71	225.55	617.57	55.69	8467.24	290.18	3.13	2199.89	0	111.62	20.71	8.69	1112.38	13124.36
4	11.71	230.27	634.86	37.66	8008.92	279.67	11.64	1932.58	0	111.62	20.71	8.40	1121.82	12411.85
3	11.71	211.09	631.60	37.66	7836.21	279.67	13.64	1970.44	9.40	111.62	20.71	11.66	1148.82	12294.22
2	11.71	211.09	634.86	37.66	7827.04	279.67	13.64	1969.49	9.40	111.62	20.71	8.40	1138.91	12274.18
1	11.71	194.61	631.60	37.66	7807.21	279.67	13.64	1990.49	25.88	111.62	20.71	11.66	1126.17	12262.62
Experiment *B*														
8	11.71	261.46	691.49	55.69	9579.55	275.37	17.95	2186.80	0	111.62	0	9.61	1112.69	14313.91
7	11.71	250.08	691.49	55.69	9343.60	275.37	17.95	2077.76	0	111.62	0	9.61	1093.52	13938.38
6	11.71	225.70	622.03	55.69	8610.29	264.58	0	2305.66	0	111.62	20.71	9.43	1061.10	13298.52
5	11.71	225.70	630.58	47.88	8578.04	290.18	3.13	2266.25	0	111.62	20.71	8.69	1074.34	13276.49
4	11.71	225.70	622.78	55.69	8498.32	293.31	0	2294.53	0	111.62	20.71	8.69	1111.62	13254.67
3	11.71	205.90	640.07	37.66	8247.89	293.31	0	2168.89	0	111.62	20.71	17.10	1021.30	12776.14
2	11.71	212.18	640.07	37.66	8225.90	293.31	0	2169.39	0	111.62	20.71	17.10	1021.30	12760.37
1	11.71	212.18	640.07	37.66	8224.33	293.31	0	2110.74	0	111.62	20.71	17.10	1055.51	12734.92
Experiment *C*														
88	11.71	261.46	691.49	55.69	9579.55	275.37	17.95	2186.80	0	111.62	0	9.61	1112.69	14313.91

(continued)

Table 3 (continued)

ROS_{max}	Year 50, Area (ha)													
(km/h)	Ct-Ec	Ct	Ec-Pb	Ec-Sb	Ec	Pb-Ec	Pb-Sb	Pb	Qr	Rp	Sb-Ec	Sb-Pb	Sb	Total
7	11.71	261.46	691.49	55.69	9419.63	278.49	14.82	2115.58	0	111.62	0	9.61	1087.69	14057.77
6	11.71	230.80	656.28	55.69	8690.20	290.18	3.13	2258.03	0	111.62	20.71	0	1093.69	13422.03
5	11.71	236.93	656.28	55.69	8698.77	261.45	3.13	2285.10	0	111.62	20.71	0	1060.14	13401.52
4	11.71	236.93	656.28	55.69	8652.68	261.45	3.13	2221.23	0	111.62	20.71	0	1134.39	13365.81
3	11.71	241.80	660.84	50.39	8568.99	280.81	12.50	2218.33	0	111.62	20.71	0.74	1107.03	13285.48
2	11.71	241.80	646.68	55.69	8529.05	280.81	12.50	2207.30	0	111.62	20.71	9.61	1058.87	13186.35
1	11.71	247.72	646.68	55.69	8451.14	280.81	12.50	2182.89	0	111.62	20.71	9.61	1029.18	13060.23

Table 4 Shrub cleaning periodicity of the optimal solution concerning all experiments. 0—no shrub cleaning; 5—shrub cleaning every 5 years; 10—shrub cleaning every 10 years

ROS_{max}	Area (ha)					
(km/ha)	0	0-10 + 10-0	10	0-5 + 5-0	5	Total
Experiment *A*						
8	13570.85	103.00	640.05	0	0	14313.91
7	12981.89	113.38	627.33	0	0	13722.59
6	12469.93	50.88	577.64	5.21	144.10	13247.75
5	12397.95	96.21	520.38	0	109.82	13124.36
4	11559.39	178.33	674.13	0	0	12411.85
3	11439.50	177.51	677.21	0	0	12294.22
2	11300.45	108.43	865.30	0	0	12274.18
1	11258.88	176.36	827.38	0	0	12262.62
Experiment *B*						
8	13570.85	103.00	640.05	0	0	14313.91
7	13252.76	84.79	600.83	0	0	13938.38
6	12592.86	60.86	472.82	0	171.97	13298.52
5	12549.50	66.91	518.02	0	142.05	13276.49
4	12430.44	43.85	575.49	0	204.89	13254.67
3	12123.86	91.44	452.57	0	108.28	12776.13
2	12148.73	68.36	460.63	0	82.65	12760.37
1	12083.94	111.95	456.39	0	82.65	12734.92
Experiment *C*						
8	13570.85	103.00	640.05	0	0	14313.91
7	13236.06	135.86	657.82	0	28.03	14057.77
6	12689.77	116.39	456.68	5.21	153.99	13422.03
5	12503.59	100.38	634.15	5.21	158.19	13401.52
4	12457.30	69.59	701.27	5.21	132.45	13365.81
3	12532.80	98.86	533.92	5.21	114.60	13285.48
2	12344.25	70.19	635.94	5.21	130.76	13186.35
1	12385.03	74.89	517.17	5.21	77.93	13060.23

6 Conclusions

In this paper, we present an iterative approach that simulates fire spread to guide the optimal selection of forest management actions over time, while incorporating suppression decisions into the process.

Integrating suppression into forest management generally increases the availability of management plans that, in the absence of suppression, would otherwise be prohibited from use. This improves key metrics such as NPV and the volume of timber removed. These effects become more pronounced when the number of bases increases to two. In addition, biodiversity, and erosion and carbon storage in general, also improve with the addition of bases. Integration of suppression reduces NPV losses compared to scenarios without fire considerations, especially with two bases, and leads to shifts in land use at the end of the planning horizon, increasing eucalyptus and pine areas. Suppression also promotes areas without shrub clearing. Greater availability of management plans that are more attractive in terms of NPV typically favors species such as eucalyptus and pine, and often excludes shrub clearing costs.

One relevant aspect of the proposed method is the resolution used to estimate fire spread. High resolutions may lead to excessive computational costs. Two straightforward ideas to overcome the computational weight of the approach are: (i) partitioning the study area in smaller regions and (ii) not considering all nodes as potential ignition locations. In the former idea, the partitioning can be based in some practical rationale (e.g. a river, as in the case study presented in this paper) or achieved through graph partitioning.

Future work could further explore the implications of varying suppression capacity (e.g. number of available bases), land use policies (e.g. prescriptions), and climate change scenarios (e.g. wind intensity) to refine forest and fuel management strategies and ensure their adaptability to evolving challenges.

Acknowledgements We thank Susete Marques (School of Agriculture, University of Lisbon, Portugal) for the forestry data about Paiva and Entre-Douro and Sousa ZIF. This work has been supported by FCT—Fundação para a Ciência e Tecnologia within the R&D Unit Project Scope UID/00319/Centro ALGORITMI (ALGORITMI/UM) and Center for Mathematical Studies, University of Lisbon (CEMS.UL)—UID/04561/2025, and within project PCIF/GRF/0141/2019 O3F—An Optimization Framework to Reduce Forest Fire, DOI 10.54499/PCIF/GRF/0141/2019.

References

1. Omi, P.N.: Theory and practice of wildland fuels management. Curr. For. Rep. **1**, 100–117 (2015). https://doi.org/10.1007/s40725-015-0013-9
2. Minas, J., Hearne, J., Martell, D.: An integrated optimization model for fuel management and fire suppression preparedness planning. Ann. Oper. Res. Springer **232**(1), 201–215 (2015). https://doi.org/10.1007/s10479-012-1298-8
3. Mercer, D.E., Haight, R.G., Prestemon, J.P.: Analyzing trade-offs between fuels management, suppression, and damages from wildfire. In: Holmes, T.P., Prestemon, J.P., Abt, K.L. (eds.) The Economics of Forest Disturbance: Wildfires, Storms, and Invasive Species, pp. 247–272. Springer Science (2008). https://doi.org/10.1007/978-1-4020-4370-3_13
4. Wei, Y.: Optimize landscape fuel treatment locations to create control opportunities for future fires. Can. J. For. Res. **42**(6), 1002–1014 (2012). https://doi.org/10.1139/x2012-051
5. Alvelos, F., Martins, I., Marques, S.: Forest management with fire simulation. Eur. J. Oper. Res. **321**(3), 991–1003 (2025). https://doi.org/10.1016/j.ejor.2024.10.013
6. Granda, B., León, J., Vitoriano, B., Hearne, J.: Decision support models and methodologies for fire suppression. Fire **6**, 37 (2023). https://doi.org/10.3390/fire6020037
7. Finney, M.A.: Fire growth using minimum travel time methods. Can. J. For. Res. **32**(8), 1420–1424 (2002). https://doi.org/10.1139/x02-068

8. Rothermel, R.C.: A mathematical model for predicting fire spread in wildland fuels. Intermountain Forest & Range Experiment Station (Vol. 115), Forest Service, US Department of Agriculture (1972). https://www.fs.usda.gov/treesearch/pubs/32533
9. Andrews, P.: Current status and future needs of the behaveplus fire modeling system. Int. J. Wildland Fire **23**(1), 21–33 (2014). https://doi.org/10.1071/WF12167
10. Fernandes, P., Gonçalves, H., Loureiro, C., Fernandes, M., Costa, T., Cruz, M., Botelho, H.: Modelos de combustível florestal para Portugal. In Actas do 6º Congresso Florestal Nacional, pp. 348–354. SPCF, Lisboa, Portugal (2009)
11. Fernandes, P., Loureiro, C.: Modelos de combustível florestal para Portugal, Documento de referência [White paper]. Departamento de Ciências Florestais e Arquitetura Paisagista, UTAD, Portugal (2021)
12. Alvelos, F.: pyo3f - a python framework for wildfire-related optimization, part i: Design and fundamentals. In: Alvelos, F., Martins, I., Rocha, A. (eds.) Advances in Optimization and Wildfire: Proceedings of the First Optimization and Wildfire Conference, October 1–4 2024, Springer, Luso, Portugal (2025).
13. Alvelos, F.: pyo3f - a python framework for wildfire-related optimization, part ii: Usage. In: Alvelos, F., Martins, I., Rocha, A. (eds.) Advances in Optimization and Wildfire: Proceedings of the First Optimization and Wildfire Conference, October 1-4 2024, Springer, Luso, Portugal. Accepted (2025)

Modelling Wind Behaviour for the Development of Scenarios in the Context of Wildfire Spread

Helena Alvelos, Ana Raquel Xambre, Francisco Marques, Agostinho Agra, and Filipe Alvelos

Abstract Wildfires cause significant losses in both material value and human lives and are becoming a frequent problem, also due to climate change. Preventing their occurrence or reducing their effect is a crucial issue. This work is part of the development of a framework that uses optimisation models for the prepositioning of resources, and for resource movement during the suppression phase. These problems involve several sources of uncertainty, one of which is the behaviour of the wind (direction and speed). It is then of the utmost importance to obtain statistical information about wind speed and direction. The aim of this work is to model historical wind data of a weather station located in a case study region in the North of Portugal (Luzim) and, for that purpose, several statistical techniques were used (e.g. descriptive statistics, correlation analysis, goodness of fit tests) in order to better understand it. It was decided to use the empirical distributions of wind speed and wind direction instead of fitting theoretical distributions, like the Weilbull Distribution used in the literature, as there is a large quantity of data available. For this, the data

H. Alvelos (✉) · A. R. Xambre
Center for Research & Development in Mathematics and Applications and Department of Economics, Management, Industrial Engineering and Tourism, University of Aveiro, Aveiro, Portugal
e-mail: helena.alvelos@ua.pt

A. R. Xambre
e-mail: raquelx@ua.pt

F. Marques
Department of Mathematics and Department of Economics, Management, Industrial Engineering and Tourism, University of Aveiro, Aveiro, Portugal
e-mail: franciscocmarques@ua.pt

A. Agra
Department of Mathematical Sciences and Center for Mathematical Studies (CEMS.UL), Faculty of Sciences, University of Lisbon, Lisbon, Portugal
e-mail: amagra@fc.ul.pt

F. Alvelos
Department of Production and Systems/ALGORITMI Research Centre/LASI, School of Engineering, University of Minho, Braga, Portugal
e-mail: falvelos@dps.uminho.pt

F. Alvelos et al. (eds.), *Advances in Optimization and Wildfire*, Lecture Notes in Operations Research, https://doi.org/10.1007/978-3-032-03108-2_7

was split into three wildfire risk seasons: low (November, December, January and February), medium (March, April, May and October) and high (June, July, August and September). Using the data from the high risk season, the values of wind speed (5 values) and of wind direction (8 values) and their respective estimated probabilities were used in order to create scenarios that were later incorporated into the optimisation models.

Keywords Wildfire · Wind speed · Wind direction · Statistical analysis

1 Introduction

Wildfires are natural ecosystem processes, but fire can and often does pose significant threats to public safety [1], property and forest resources [2]. Mediterranean countries are very susceptible areas to wildfires [3] due to its winter precipitation with mild temperatures, alternating with summer drought under high temperatures, which in turn, promotes fuel accumulation and general flammability of the landscape [2, 4, 5]. Among the Mediterranean countries, Portugal is one of the countries most affected by wildfires, with the number of ignitions increasing in recent years [6, 7]. As observed in [8], large wildfires start in more remote areas, due to the lower population density and distance from main roads. Most of the country is described by very fragmented landscapes composed of very small land plots belonging to different landowners which increase the number of fire ignitions [6, 9]. In Portugal, one of the most recent devastating years was 2017, where there were over 3000 wildfires, resulting in over 400 000 hectares of burned area and over 110 human casualties [10]. Although extremely large wildfires are a very small minority, they are responsible for the most part of the burned area [11], and as such, the most part of the firefighting budget.

Thus, wildfire management is a significant topic that demands effort from many operatives and researchers, and, according to [12], it can be divided into four main areas: fuel management, prevention, detection and management of fire suppression resources.

Although physical weather forecasting methods provide higher accuracy, statistical methods still present several advantages over the former, such as being more computationally efficient, and able to assess the uncertainty in each location of the map [13]. Many studies relating to the statistical analysis of wind resources have been made due to the increasing need of wind power, however, there is a lack of research on the statistical modelling of wind for wildfire suppression.

In the literature it is common to find that the wind speed is described by the Weibull Distribution Function (WDF) [14, 15], but also the Gamma, Rayleigh [16] and Lognormal distributions [17]. The two-parameter WDF, which is a special case of the generalized Gamma distribution [18], is the most common application of this distribution in analysing wind speed. This distribution, while being a generally accurate wind speed representation, requires long term data to accurately fit a WDF [13].

This study focuses on modelling historical wind data from a weather station in northern Portugal, with the goal of addressing uncertainties related to wind behaviour (speed and direction). The research employs various statistical techniques to analyse wind speed and wind direction, and the estimated probabilities of these values were used to create scenarios, which were then incorporated into optimisation models.

In the following section the several analyses of the data related to wind behaviour will be described, as well as some of the main results obtained. The final section highlights the main conclusions and contributions of this work.

2 Data Analysis

For this study, the data used was the average hourly wind data provided by IPMA [1], collected at a Portuguese weather station located in Luzim, Porto, 18 km away from the case study area (Fig. 1), regarding the period between June 2012 and July 2023.

Initially, there were 119,000 data records, however, some of the data was collected with errors, due to technical malfunctions. Eliminating those records resulted in 83,400 records that were used for the subsequent statistical analyses. The variables used in the analyses were the year, month, day and hour, the maximum speed per hour, the hourly average speed and the hourly average direction.

From this data, histograms of the hourly average direction and hourly average speed of the wind were created and are presented in Fig. 2 (the wind speed values that exceed 6 m/s are grouped in one class) and Fig. 3.

It can be noticed that the average wind speed presented an asymmetric shape. It was also determined the average value, as well as the standard deviation, of the 83,400 values: 1.653 m/s 1.079 m/s, respectively.

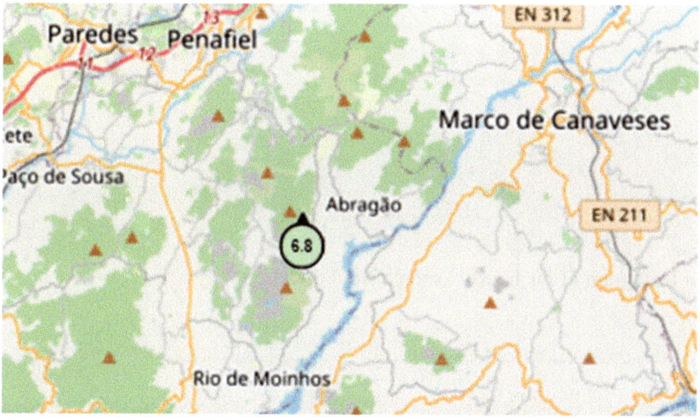

Fig. 1 Location of the weather station (Luzim)

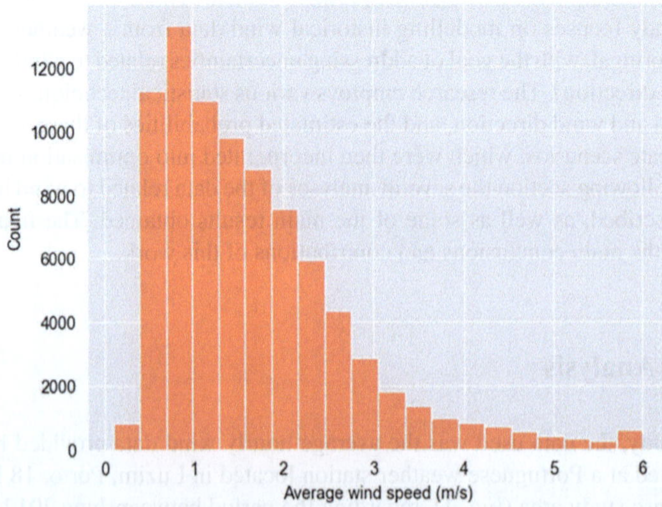

Fig. 2 Histogram of the hourly average wind speed (m/s)

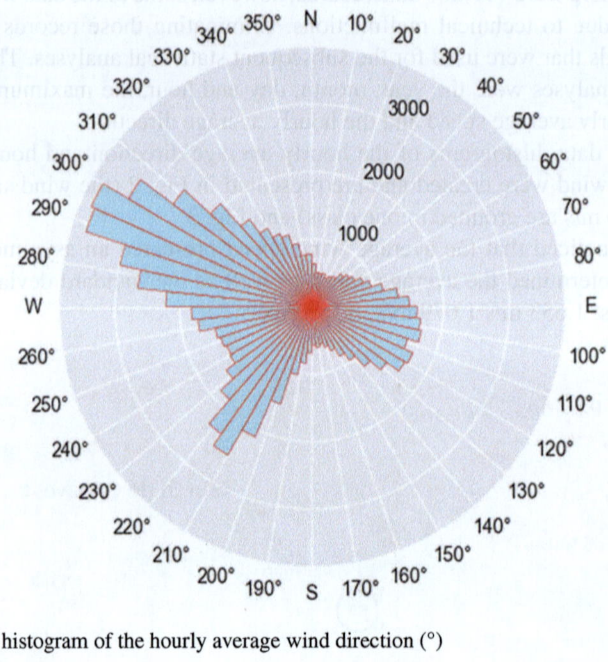

Fig. 3 Polar histogram of the hourly average wind direction (°)

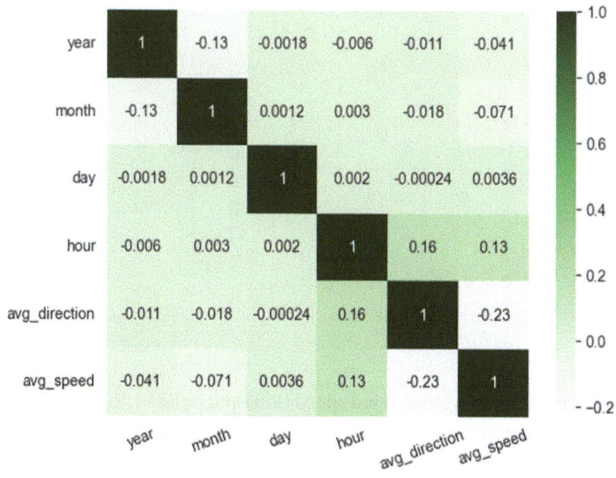

Fig. 4 Correlation matrix

Regarding the wind direction, it was predominantly WNW, around 290°–310°, and three main groups were observed: WNW, as mentioned, and also ESE and SSW.

In Fig. 4 the correlation matrix obtained for the entire set of data is presented. It shows that there is little or no correlation between wind speed and direction (−0.23), as well as between these variables and the remaining variables, namely the time variables such as the hour, day, month, and year of the record.

The correlation between these time variables shows that there are no linear relations between them.

Some univariate and multivariate analyses were performed but no helpful results were obtained.

Due to weak correlations in the full dataset, the data was segmented into three seasons, taking into account the Portuguese climate and the associated wildfire risk: low, medium and high risk.

The three groups were defined trough the analysis of the profile of the wind behaviour in each month, grouping the similar months.

2.1 Analysis of the Three Seasons

The low risk season contains data from the months of November, December, January and February, the medium risk contains the months of March, April, May and October and the high risk season contains data belonging to the hotter months of June, July, August and September.

In Figs. 5, 6, and 7 the histograms of the wind speed and direction for each risk season are presented, and in Table 1, the mean and standard deviation of the wind speed and of the wind direction, for each dataset, are shown.

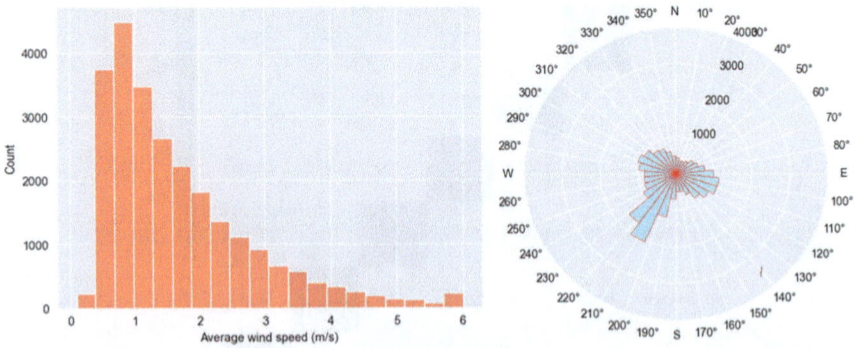

Fig. 5 Histogram of the hourly average wind speed (left) and polar histogram of the hourly average wind direction (right) during the low risk season

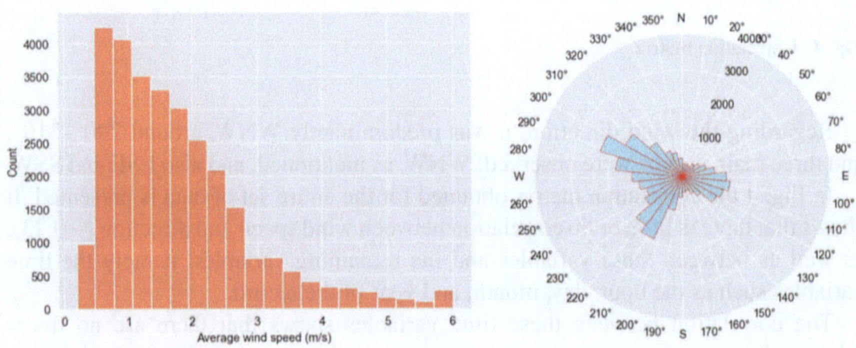

Fig. 6 Histogram of the hourly average wind speed (left) and polar histogram of the hourly average wind direction (right) during the medium risk season

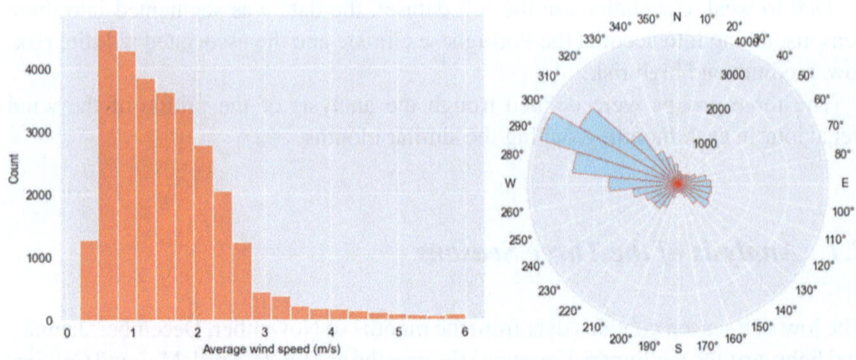

Fig. 7 Histogram of the hourly average wind speed (left) and polar histogram of the hourly average wind direction (right) during the high risk season

Table 1 Descriptive statistics for wind speed and wind direction for all data and data divided by risk season

Data	Number of records	Wind speed (m/s)		Wind direction (°)	
		Mean	Standard deviation	Mean	Standard deviation
All data	83,425	1.653	1.079	210.208	93.746
Low risk	24,993	1.656	1.161	191.926	91.942
Medium risk	29,382	1.759	1.137	206.082	92.916
High risk	29,050	1.544	0.924	230.109	92.375

Table 2 Correlation index between wind direction and hour of the day, for the three seasons

Season	Low risk	Medium risk	High risk
Correlation index (r)	0.082	0.155	0.227

From this initial exploratory data analyses, and with the goal of better understanding the groups identified in the polar histograms of the wind direction (Figs. 5, 6, and 7), it was decided to use the hour of the day and determine its correlation with the wind direction (Table 2).

Although the correlation values are low, it can be verified that they grow when moving from the low risk to the high risk season, which indicates that in the high risk season the wind direction is more associated with the hour of the day, than in the other seasons. It was thus decided to concentrate the analysis on the high risk season, also because the impact of wildfires is much bigger during the summer months.

2.2 *High Risk Season*

In order to model the wind behaviour, namely the wind speed and wind direction, the data from the high risk season (29,050 records) was used. As a first approach, there was an attempt to fit theoretical distributions to the available data. As such, the *FITTER* Python module was used, which fits every distribution function from *Scipy* (another Python module) to the data, and estimates the parameters of the best fitted distribution.

The *FITTER* module found that the Beta distribution was the best fit for the wind speed, based on the sum square error. The estimated parameters are: $\alpha = 2.17366$; $\beta = 1{,}961{,}559$, $loc = 0.19094$, and $scale = 1{,}223{,}199$. Both the fitted distribution functions and the high risk season histogram are presented in Fig. 8.

A similar approach was done for the wind direction, however, it was understood that the wind direction could not be estimated by only one distribution. Therefore, eight Gaussian mixture functions were fitted, and in Fig. 9 four of the resulting histograms are presented.

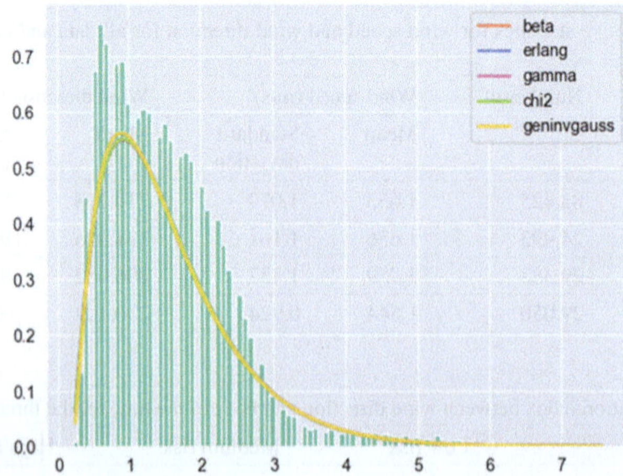

Fig. 8 Histogram of the hourly average wind speed (m/s) for the high risk season and fitted theoretical distributions (note that the curves of the fitted distributions are overlapping)

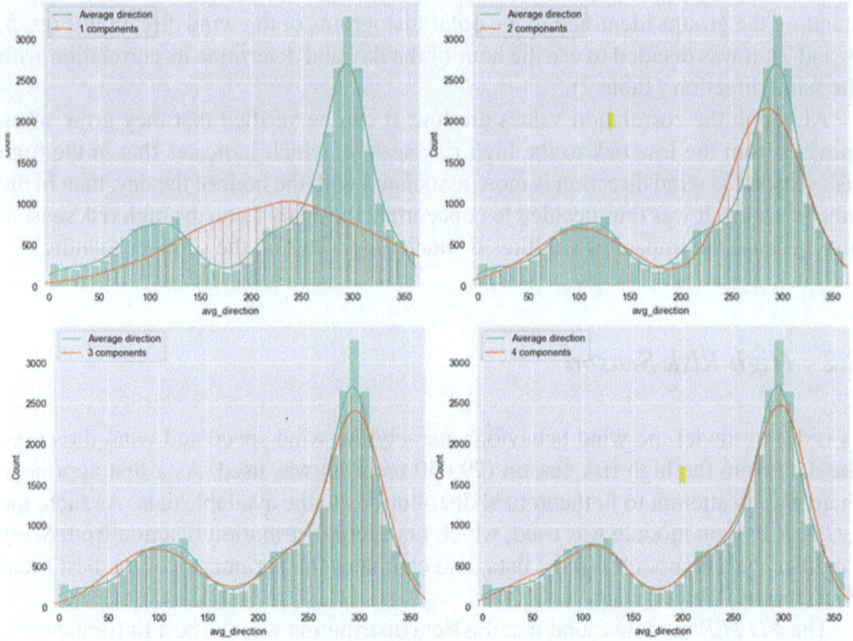

Fig. 9 Histograms of the high risk season hourly average wind direction (°) and four of the fitted Gaussian Mixture distribution functions

Considering the results, the three components Gaussian mixture function was chosen since it presents a good fit and it does not become too complex.

The estimated distribution function is:

$$0.16.\,N\left(231, 26.9^2\right) + 0.28.\,N\left(97, 43.7^2\right) + 0.56.\,N\left(297, 24.0^2\right).$$

For the identified distributions, goodness of fit tests were performed and it was noticed that the null hypothesis (that represents the adequate fit of the data to the theoretical distribution) was rejected in both cases (wind speed and wind direction). These results can be attributed to the large amount of data, which makes it difficult for the hypothesis not to be rejected, so it was decided to use the real historical data for creating the scenarios to be incorporated into the optimisation models.

2.3 Scenarios Definition

The previous study of the data from the high risk season, was essential for creating a set of scenarios to be used in the optimisation models. It should be noticed that the number of scenarios was limited so as not to increase the computational effort of the models and inhibit obtaining timely solutions, so the wind speed data was divided into 5 classes, and the wind direction was split into 8 classes. In Table 3 the number of observations within each joint class is presented, as well as the total number of observations per class of wind speed and wind direction.

By combining the wind speed classes and the wind direction classes, the resultant number of scenarios was 40, and the joint probabilities of the wind speed and wind direction were estimated using the relative frequency of each joint class (see Table 4). For each scenario, the joint probabilities were used, as well as the central values for each interval of wind speed and wind direction (also presented in Table 4).

These scenarios were combined with other uncertainty sources that influence wildfire behaviour (e.g. ignition location) and incorporated into the optimisation models used for the prepositioning of resources, and for resource movement during the suppression phase of the fire.

In the next section, some additional analyses used to better explore the data are presented.

2.4 Further Analyses

As seen previously, there is no linear relation between the wind speed and wind direction, however after plotting a polar chart, where the average wind direction is represented using the circular axis and the average wind speed is represented using the radial axis (see Fig. 10), it showed that there is some relationship between the two variables.

Table 3 Number of observations per joint class

		Wind Direction (°)								
		]0, 45]	]45, 90]	]90, 135]	]135, 180]	]180, 225]	]225, 270]	]270, 315]	]315, 360]	Total
Wind Speed (m/s)	]0, 1.5]	731	745	1816	921	1277	3101	5775	2110	16,476
	]1.5, 3.0]	373	1031	924	417	693	973	5566	1068	11,045
	]3.0, 4.5]	54	271	426	45	75	14	115	81	1081
	]4.5, 6.0]	2	151	232	5	7	0	1	0	398
	]6.0, 8.0]	1	15	34	0	0	0	0	0	50
	Total	1161	2213	3432	1388	2052	4088	11,457	3259	29,050

Table 4 Estimated joint probability of wind speed and wind direction during high risk season

Central Values	22.5°	67.5°	112.5°	157.5°	202.5°	247.5°	292.5°	337.5°
0.75 m/s	0.0252	0.0257	0.0625	0.0317	0.0440	0.1068	0.1988	0.0726
2.25 m/s	0.0128	0.0355	0.0318	0.0144	0.0239	0.0335	0.1916	0.0368
3.75 m/s	0.0019	0.0093	0.0147	0.0016	0.0026	0.0005	0.0040	0.0028
5.25 m/s	0.0001	0.0052	0.0080	0.0002	0.0002	0.0000	0.0000	0.0000
7.00 m/s	0.0000	0.0005	0.0012	0.0000	0.0000	0.0000	0.0000	0.0000

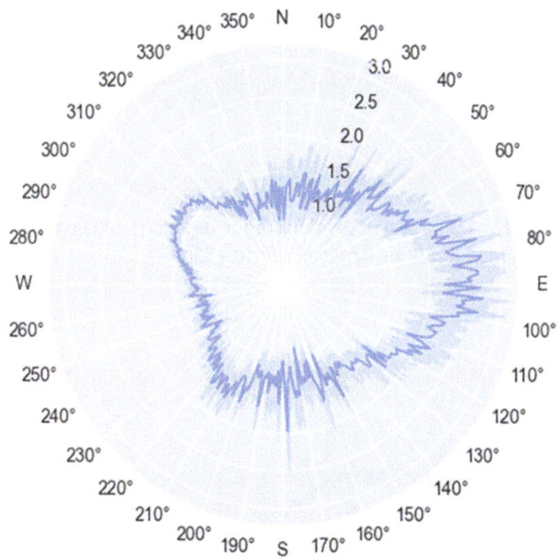

Fig. 10 Polar plot of the hourly average wind direction and the hourly average wind speed

It can be observed that wind directions between 45° and 135° have high values of average wind speed and present higher variability. The opposite is verified between the directions 270° (W) and 315°, where the wind speed is lower and there is less variability.

By further analysing the data from these two wind direction intervals, it was detected that the higher speed variability period occurs most commonly during the morning, and the lower variability period during the late afternoon and night.

For the first wind direction interval, 45° to 135°, there were a total of 5672 observations, which represent 20% of the total number of observations (29,050). From those, 64% of the data is observed between 7 h-12 h (see Fig. 11, left), and has a high average speed with high variability and outliers (see Fig. 11, right).

As for the second wind direction interval, 270° to 315°, the total number of observations was 11,606, which represent 40% of the total number of observations (29,050). From those, 53% of were observed between 16 h-23 h (see Fig. 12, left). It can be noticed that, for that interval, the average speed is lower, with lower variability and less outliers (see Fig. 12, right).

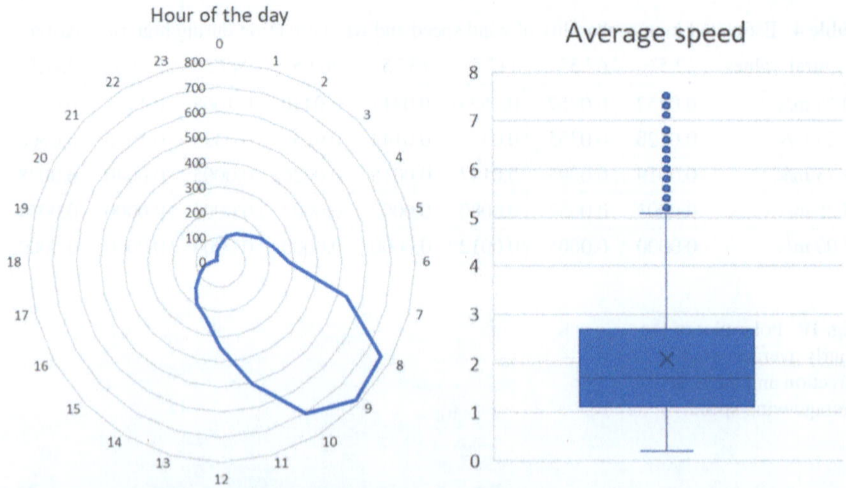

Fig. 11 Polar plot of the hour of the day (left) and boxplot of the hourly average wind speed (right) for wind directions between 45° to 135°

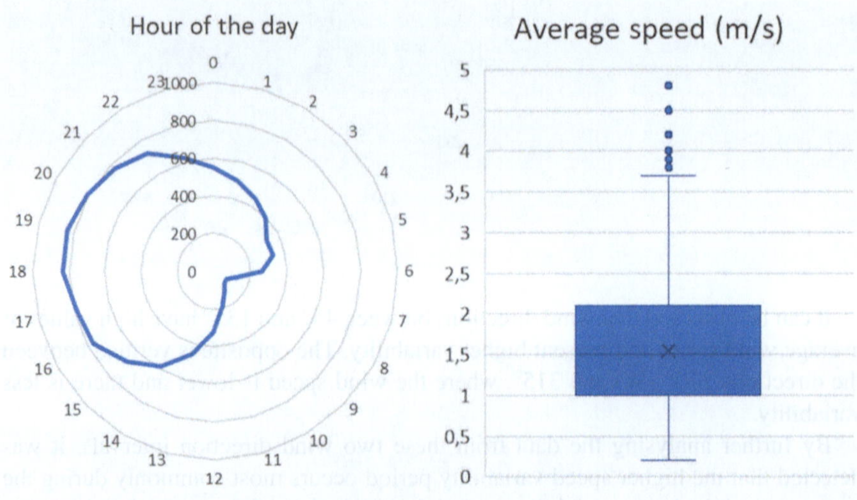

Fig. 12 Polar plot of the hour of the day (left) and boxplot of the hourly average wind speed (right) for wind directions between 270° to 315°

From this final analysis it was possible to understand that the hour of the day also can be a factor when modelling wind speed and direction.

3 Conclusions

This study was based on a large set of real data and shows that it is possible to better understand wind behaviour when the data represents accurately what happens in the case study location. Also, it was possible to treat a large number of records, eliminating corrupted records and records with missing values, and unveiling some relations between the variables studied. That led to the identification of a high risk season, which data was further analysed.

It was not possible to obtain theoretical distributions that adequately represent the behaviour of the variables, however it could be noticed that when there is a large amount of historical data it is better to use empirical distributions derived from it.

Nonetheless, it was possible to estimate the joint distribution of the wind speed and wind direction in the case study area and develop the required scenarios. Those scenarios were included into the framework that uses optimisation models for the prepositioning of firefighting resources, and for resource movement during the suppression phase of wildfires.

Also, a relationship was found between the wind speed and wind direction according to the hour of the day. As future work, it is important to further study this relationship and estimate joint probabilities according to the time of the day.

It was not possible to integrate, at this time, those findings into the framework, however, in future work new scenarios should be considered that include the hour of the day, as well as all the other variables already incorporated.

Additionally, since the analysis was based on the data from one weather station, it would be interesting to replicate the study with data from other stations, in order to determine if the findings are similar, as well as identify the differences between the data from several stations.

This work is important to help in better understanding wind behaviour, and consequently fire spreading behaviour, thus contributing to a crucial issue, namely wildfire prevention and suppression.

Acknowledgements This work was supported by the Portuguese Foundation for Science and Technology (FCT) within the R&D Units Project Scope UIDB/00319/2020 and within project PCIF/GRF/0141/2019 "O3F—An Optimization Framework to Reduce Forest Fire" https://doi.org/10.54499/PCIF/GRF/0141/2019.

This work was also supported by CIDMA under the Portuguese Foundation for Science and Technology (FCT, https://ror.org/00snfqn58) Multi-Annual Financing Program for R&D Units, grants UID/4106/2025 and UID/PRR/4106/2025.

References

1. Miranda, A.I., Amorim, J.H., Martins, V., Pimentel, C., Rodrigues, R., Tavares, R., Borrego, C.: Numerical modelling of the impact of wildland-urban interface fires on Coimbra air quality. WIT Trans. Ecol. Environ. **119**, 333–342, ISSN: 17433541 (2008). https://doi.org/10.2495/FIVA080331

2. Keeley, J.E., Bond, W.J., Bradstock, R.A., Pausas, J.G., Rundel, P.W.: Fire in Mediterranean Ecosystems: Ecology, Evolution and Management. Cambridge University Press, vol. 9780521824910, pp. 1–515, ISBN: 9781139033091 (Jan. 2011). https://doi.org/10.1017/CBO9781139033091
3. Pausas, J.G., Fernández-Muñoz, S.: Fire regime changes in the western Mediterranean basin: From fuel-limited to drought-driven fire regime. Clim. Change. **110**, 215–226, ISSN: 01650009 (1–2 Jan. 2012). https://doi.org/10.1007/s10584-011-0060-6
4. Moreira, F., Vaz, P., Catry, F., Silva, J. S.: Regional variations in wildfire susceptibility of land-cover types in Portugal: implications for landscape management to minimize fire hazard. Int. J. Wildland Fire. **18**, 563–574, ISSN: 10498001 (5 2009). https://doi.org/10.1071/WF07098
5. Viedma, O., Moreno, J.M., Rieiro, I.: Interactions between land use/land cover change, forest fires and landscape structure in Sierra de Gredos (Central Spain). Environ. Conserv. **33**, 212–222, ISSN: 03768929 (3 Sep. 2006). https://doi.org/10.1017/S0376892906003122
6. Silva, J.S., Vaz, P., Moreira, F., Catry, F., Rego, F.C.: Wildfires as a major driver of landscape dynamics in three fire-prone areas of Portugal. Landsc. Urban Plan. **101**, 349–358, ISSN: 01692046 (4 2011). https://doi.org/10.1016/j.landurbplan.2011.03.001
7. Bento, V.A., Lima, D.C., Santos, L.C., Lima, M.M., Russo, A., Nunes, S.A., DaCamara, C.C., Trigo, R.M., Soares, P.M.: The future of extreme meteorological fire danger under climate change scenarios for Iberia. Weather. Clim. Extrem. **42**, ISSN: 22120947 (Dec. 2023). https://doi.org/10.1016/j.wace.2023.100623
8. Catry, F.X., Rego, F.C., Moreira, F., Bacao, F.: Characterizing and modelling the spatial patterns of wildfire ignitions in Portugal: fire initiation and resulting burned area. WIT Trans. Ecol. Environ. **119**, 213–221, ISSN: 17433541 (2008). https://doi.org/10.2495/FIVA080221
9. Gaspar, J., Fidalgo, B.: Evolução do uso solo e avaliação do valor paisagístico e de recreio na área de paisagem protegida da Serra do Açor. [Online]. https://www.researchgate.net/publication/237777442 (2014)
10. Departamento de Gestão de Áreas Públicas e de Proteção Florestal.: 9º relatório provisório de incêndios florestais—2017: 01 de janeiro a 16 de outubro. (2017)
11. Strauss, D., Bednar, L., Mees, R.: Do one percent of forest fires cause ninety-nine percent of the damage? pp. 319–328. [Online]. https://academic.oup.com/forestscience/article/35/2/319/4642523. (1989)
12. Martell, D.L.: A review of recent forest and wildland fire management decision support systems research. (Jun. 2015). https://doi.org/10.1007/s40725-015-0011-y
13. Veronesi, F., Grassi, S., Raubal, M.: Statistical learning approach for wind resource assessment. (Apr. 2016). https://doi.org/10.1016/j.rser.2015.11.099
14. Salim, O. M., Dorrah, H. T., Hassan, M. A.: Wind speed estimation based on a novel multivariate Weibull distribution. IET Renew. Power Gener. **13**, 2762–2773, ISSN: 17521424 (15 Nov. 2019). https://doi.org/10.1049/iet-rpg.2019.0313
15. MacEira, M.E., Melo, A.C., Pessanha, J.F., Cruz, C.B., Almeida, V.A., Justino, T.C.: Wind uncertainty modeling in long-term operation planning of hydro-dominated systems. Inst. Electr. Electron. Eng. Inc. ISBN: 9781665412117 (2022). https://doi.org/10.1109/PMAPS53380.2022.9810576
16. Ali, S., Lee, S.M., Jang, C.M.: Statistical analysis of wind characteristics using Weibull and Rayleigh distributions in Deokjeok-do Island—Incheon, South Korea. Renew. Energy. **123**, 652–663, ISSN: 18790682, (Aug. 2018). https://doi.org/10.1016/j.renene.2018.02.087
17. Dong, Z., Wang, X., Zhu, R., Dong, X., Ai, X.: Improving the accuracy of wind speed statistical analysis and wind energy utilization in the Ningxia autonomous region, China. Appl. Energy. **320**, ISSN: 03062619, (Aug. 2022). https://doi.org/10.1016/j.apenergy.2022.119256
18. Agarwal, S.K., Kalla, S.L.: A generalized gamma distribution and its application in reliability. Commun. Statistics—Theory Methods. **25**, 201–210, ISSN: 03610926 (1 1996). https://doi.org/10.1080/03610929608831688

Forest Road Network for Firefighter Access

Miguel Constantino, Marta Mesquita, and Mariana Sempiterno

Abstract Forest fires pose a growing threat to the environment, economy, and human life, demanding immediate prevention and firefighting strategies. Despite the vital role of forest roads in granting access to firefighters, their planning frequently overlooks firefighting requirements. In this paper, we discuss models designed to identify the most cost-effective road network for ensuring firefighter access to the forest. We model the problem as a Steiner Tree Problem in a graph and present results from a case study conducted in a Portuguese forest.

Keywords Wildfire management · Forest access road optimization · Steiner tree problem

1 Introduction

Forest fires are considered one of the most significant threats to forest resources globally, particularly in Mediterranean countries, given the prevalence of hot, dry summers [23]. These fires present a significant challenge to the environment, the economy and, in extreme cases, human life. In addition to destroying existing biodiversity and jeopardising the sustainability of the forest, they contribute to greenhouse

M. Constantino (✉)
Center for Mathematical Studies and Faculdade de Ciências da Universidade de Lisboa, Lisboa, Portugal
e-mail: mfconstantino@ciencias.ulisboa.pt

M. Mesquita
Center for Mathematical Studies and Instituto Superior de Agronomia, Universidade de Lisboa, Lisboa, Portugal
e-mail: martaoliv@isa.ulisboa.pt

M. Sempiterno
Faculdade de Ciências da Universidade de Lisboa, Lisboa, Portugal
e-mail: mariana_sempiterno@hotmail.com

F. Alvelos et al. (eds.), *Advances in Optimization and Wildfire*, Lecture Notes in Operations Research, https://doi.org/10.1007/978-3-032-03108-2_8

effects on a large scale due to the gases emitted [3]. The recent increase in temperature and the occurrence of abnormally hot periods have extended the fire season, with projections indicating a further extension in the coming years [14].

The subject of forest resource management has been the focus of study for many decades. This has encompassed the question of which areas should be harvested, the timing of harvest, the location of wood gathering and the routes for wood transportation. However, the introduction of the fire protection component is a relatively recent development [7]. The objective of forest protection management is to mitigate the impact of forest fires. This can be achieved through a variety of strategies, including the prevention of fires caused by unnatural factors (such as burning or arson), the enhancement of fire detection capabilities, and the facilitation of timely intervention and response.

In the context of forest fires, access to the site is often challenging due to the lower density of roads compared to urban areas. One strategy for enhancing firefighter intervention is to guarantee accessibility to forest regions, as an area's vulnerability to fire is dependent, among other factors, on its distance from roads [24]. Forest roads serve as the fundamental support infrastructure for firefighting and fire prevention. All ground-based operations rely on them, underscoring the importance of well-designed road planning [21]. However, the majority of studies examining the construction of these roads do not consider their role in forest protection [12]. Forest roads are not only pivotal for firefighting but also for identifying fires, enabling more effective monitoring, particularly in areas prone to fires [21].

Several authors have addressed the problem of building forest roads to facilitate the access of areas of forest in order to improve firefighting abilities.

In the event of a fire, firefighters need to be able to reach the scene as quickly as possible. This is why one of the most studied problems involves determining the shortest route between fire stations or lookout points and forests [23]. However, it is not always possible to use the shortest route due to the constraints that can arise in the event of a fire. Accordingly, Akay et al. [3] sought to find the best route by defining it not only as the fastest but also as the safest or most reliable.

In forest road construction, the problem is generally considered to be multi-objective [13, 16, 25]. Stefanovic et al. [21] sought to balance a low cost, a short distance to logistics centers, and a road density that allowed accessibility.

The notion of a protected zone (corridors that extend along roads and which it is considered possible to defend in the event of a fire) has been addressed by multiple authors. For Psilovikos et al. [17] there are two concepts of protected zone: the theoretical one, with 500 meters on downhill slopes and 300 meters on uphill slopes, and the practical one, a little smaller, with 300 and 150 meters respectively. However, these measures were defined for a non-productive forest. Stefanovic et al. [21] defined 100 meters as the maximum limit since this is the length of the hoses found in local firefighting units (Serbia). On the other hand, in China the available hoses allow up to 200 meters to be defined as a safe zone on flat terrain [25]. These distances were also studied by Campbell et al. [4] who analyzed the various escape routes for firefighters in the event of an emergency and studied pedestrian travel times in different types of terrain and vegetation. Sullivan et al. [22] went further and carried out the same studies (with different slopes and vegetation) but analyzed firefighters with all their equipment on to consider the weight they carry in real situations.

Roads have also been studied for their potential as a barrier to the progression of fires. Barriers to the advance of fires can be natural, such as corridors with low-flammable vegetation (fuel breaks) and watercourses in the area, or artificial, such as areas cleared of any vegetation (fire breaks). Zong et al. [26] concluded that the density of barriers in a forest has a greater impact than the width of any barrier and that their spatial pattern also influences their effectiveness. However, roads have an almost negligible effect in this case study since the barriers identified as most effective were at least 30 meters wide. The potential of forest roads in protecting the forest lies in allowing firefighters access and not so much in serving as a barrier to wildfires.

Another definition of a protected area is determined by the time it takes firefighters to get from their usual location (fire station or lookout point) to each point in the forest, with a consensus around the maximum 15/20 minutes needed to effectively fight fires [3, 17]. Some studies define the maximum access time according to the number of fires in recent years; the higher the number, the shorter the time needed to reach the location [2].

In 1998, Murray [15] presented a formulation for the problem called the Multiple Target Access Problem, i.e. defining the best road to build in order to guarantee access to multiple points, in this case the points where the wood that had already been cut would be collected. A mixed integer programming formulation was used, defining an oriented flow problem guaranteeing that all the wood reached the connection points to the existing road network.

In this paper, we consider a close problem, the Network Covering Problem (*NCP*), which is defined formally in next section. In a forest, a set of sites must be *accessible* meaning they should be within the reach of firefighters via road access. The forest may already have some existing roads and there is a set of potential forest roads that can be built. The objective is to build the necessary roads so that the required accessibility level is achieved at minimum cost.

The difference to the Multiple Target Access Problem is that in this problem the roads must pass by or reach a set of predefined points, while in the *NCP* the roads should *allow the access* to a set of predefined points. This access can be obtained by proximity and more than one road may be able to provide it. So *NCP* is more general than the Multiple Target Access Problem.

In next section we define the Network Covering Problem and show how it can be modeled as a Steiner Tree Problem in a graph. In Sect. 3 we present the case study of a forest in the north of Portugal. In Sect. 4 we present results for solving several problem instances from the case study, with a state-of-the art Steiner Tree Problem solver. Finally in last section we discuss the results and present possible future extensions of this work.

2 Problem and Model

To define the Network Covering Problem (*NCP*), we consider a forest with a selected set of sites that must be accessible. Each site may represent an area for which it is the geographical center with known coordinates. The forest contains a set of existing roads, and there is also a set of potential forest roads that can be built. We assume that each forest road consists of segments (edges) defined by their endpoints (nodes), whose coordinates are known. The sets of edges and nodes are denoted by *EF* and *NF*, respectively. The set of nodes in *NF* that also lie on an existing road is denoted by *NR*. Forest roads can be built starting from nodes in *NR*.

In this paper, we assume that a site is accessible if its distance from a road does not exceed a predetermined value. The road can be either an existing road or a built forest road. Sites already accessible from existing roads may be discarded from the problem. Let *S* denote the set of sites that are not accessible by an existing road.

The aim is to construct the necessary forest roads with edges in *EF* such that each site $i \in S$ is accessible. Furthermore, each forest road must connect to the existing road network in a node of *NR*. The total cost of building roads should be minimized.

A site $i \in S$ is accessible if it is within a maximum distance $D_{\max}$ from a road. We consider a distance function $D(a, b)$ defined for any two forest points *a* and *b*. For instance, the *Euclidean distance* can be used, $D(a, b) = \sqrt{(a_x - b_x)^2 + (a_y - b_y)^2}$, where (a_x, a_y) and (b_x, b_y) are the coordinates of points *a* and *b*, respectively.

Using a distance function *D*, we define accessibility as follows: a site $i \in S$ is accessible if there exists a node $c \in \text{NF}$, in a built road, such that $D(c, i) \leq D_{\max}$. In this case, we say that node *c* covers site *i* (Fig. 1 i).

Although this may be the simplest way to define accessibility with a distance function, in some cases it may not be accurate. Even if the nodes of an edge are farther than $D_{\max}$ from site *i*, some points along the edge may still lie within that distance (Fig. 1 ii).

To address this limitation, consider an edge $\{a, b\}$ with endpoints *a* and *b*. If there is a point *c* on the edge within the required distance, the edge $\{a, b\}$ can be replaced by two smaller edges, $\{a, c\}$ and $\{c, b\}$ (Fig. 1 iii).

This replacement has another advantage. If $\{a, b\}$ is the last segment of a road starting at a node in *NR*, it may not need to be built entirely. It is sufficient to build only the segment $\{a, c\}$ or $\{c, b\}$ that ends at *c*. Taking this idea further, the segment $\{a, b\}$ can be split at two points, c_1 and c_2, that are exactly at a distance $D_{\max}$ from site *i*, and replaced by three smaller segments: $\{a, c_1\}$, $\{c_1, c_2\}$, and $\{c_2, b\}$. Thus, if $\{a, b\}$ is the last segment of a road, only one of the smaller segments, $\{a, c_1\}$ or $\{c_2, b\}$, needs to be built (Fig. 1 iv).

By applying the aforementioned changes, the problem can be modeled as covering sites with nodes on roads. In other words, roads should be built such that for each $i \in S$, there exists a node in *NF* at a distance from *i* not exceeding $D_{\max}$. An example of a solution is shown in Fig. 2.

If the minimum distance from a site $i \in S$ to the closest node in *NF* exceeds $D_{\max}$, the problem becomes infeasible. If removing *i* from the problem is not an option, we

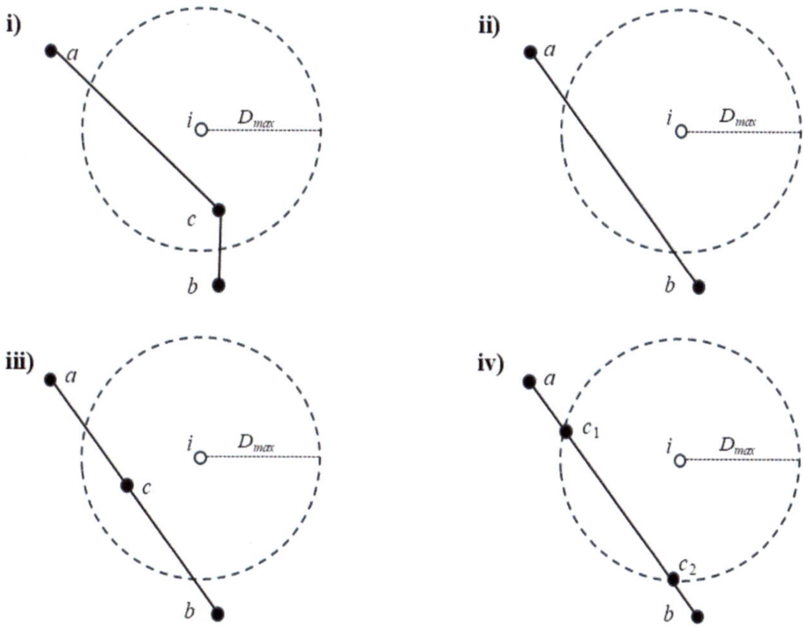

Fig. 1 This figure illustrates accessibility to site \$i\$: (i) node c covers site i as $D(c, i) \leq D_{\max}$; (ii) nodes a and b do not cover i but there are points in the edge $\{a, b\}$ within the distance $D_{\max}$; (iii) c is in a edge within $D_{\max}$ so $\{a, b\}$ can be replaced by two smaller edges, $\{a, c\}$ and $\{c, b\}$; iv) $\{a, b\}$ can be replaced by $\{a, c_1\}$, $\{c_1, c_2\}$, and $\{c_2, b\}$.

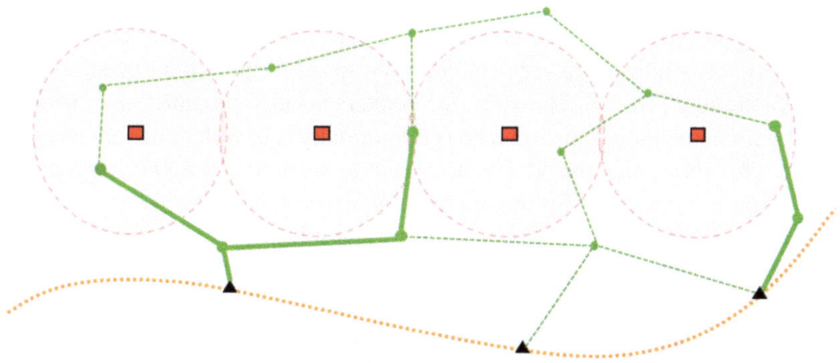

Fig. 2 An example of *NCP*. Sites in *S* are represented by red squares, forest road segments (edges in *EF*) and their endpoints (nodes in *NF*) are green, an existing road is displayed in orange and nodes in *NR* are represented by black triangles. A solution is shown in bold green. Circles around red squares define the access area for sites in *S*

may instead consider a larger accessibility distance at a cost, which corresponds to the necessity of building an unplanned road to access the site.

Formally, the Network Covering Problem, *NCP*, can be defined as the problem of finding a minimum cost road network covering all sites in *S*. Next we show how to model the *NCP* as a Steiner Tree Problem (*STP*) [10].

To define the *STP* let $G = (V, E)$ be a graph where V is the set of nodes and E is the set of edges (unordered pairs of nodes). A subgraph of G is a graph $H = (V_0, E_0)$, with $V_0 \subseteq V$ and $E_0 \subseteq E$. A tree in G is a connected subgraph without cycles [6].

Given $G = (V, E)$ with costs associated to edges, and a set of nodes $T \subseteq V$ called *terminal nodes*, the *STP* consists in finding a tree in G of minimum cost connecting the nodes of T. That is, a minimum cost subgraph $H = (V_0, E_0)$ which is a tree and $T \subseteq V_0$. Observe that if the edge costs are nonnegative, the condition of H not having cycles does not need to be imposed explicitly. If there were a cycle, one of its edges could be removed and the subgraph would remain connected without increase in the cost.

Now we define a graph $G = (V, E)$ for which solving the Steiner Tree Problem also solves *NCP*. The set of nodes V is composed by the nodes in *NF*, the sites in *S* and an extra node s_0, that is $V = \text{NF} \cup S \cup \{s_0\}$. The terminal node set is $T = S \cup \{s_0\}$. The set of edges E is composed by the edges in *EF*, edges linking s_0 to nodes in *NR*, that is, in $\text{ER} = \{\{s_0, a\}, a \in \text{NR}\}$, and edges representing the accessibility from nodes in *NF* to sites in *S*, that is, in $\text{ES} = \{\{a, i\}, a \in \text{NF}, i \in S : i \text{ is accessible from } a\}$. Then $E = \text{EF} \cup \text{ER} \cup \text{ES}$. In *STP*, costs for edges in *EF* are the same as the corresponding costs in *NCP*; costs for edges in *ER* are zero; costs for edges in *ES* are a big number, to prevent terminal nodes to be visited twice, added by the accessibility cost in the case of sites farther than $D_{\max}$ to the nearest node in *NF* (Fig. 3).

Modeling the *NCP* as the *STP* is of practical interest because the latter problem has been extensively studied, and very efficient solution algorithms have been developed by several authors [10, 19]. However, the converse is also possible: any instance of the *STP* with nonnegative costs can be (polynomially) converted into an instance of the *NCP*. This shows that the *NCP* is not easier to solve than the *STP* in general and its resolution is a challenge for the operational research community.

An instance of the *STP* is characterized by a graph $G = (V, E)$, a non-empty set $T \subseteq V$ of terminal nodes, and a cost function c_e defined on the edges $e \in E$. To define an equivalent instance of the *NCP*, we need to specify the set of sites *S*, the set of nodes *NF* and $\text{NR} \subseteq \text{NF}$, the set of edges *EF* linking nodes in *NF*, and, for each site in *S*, the set of nodes that can access it.

Let $\text{NF} = V$ and $\text{EF} = E$. Choose any $t \in T$ and set $\text{NR} = \{t\}$. Define *S* as a duplicate of $T \setminus \{t\}$, where each corresponding original node is already in *NF*. $\text{ES} = \{(a, s) : a \in T \setminus \{t\}, s \in S\}$ Each site in *S* is accessed only by the corresponding original node in $T \setminus \{t\}$. For $e \in \text{EF}$, set $\text{cost}(e) = c_e$. $e \in \text{ES}$ $\text{cost}(e) = 0$.

Since any node in *S* is only accessible through the corresponding node in $T \setminus \{t\}$, all these nodes must be included in a solution to the *NCP*. As node *t* is the only node in *NR*, it must also be included in that solution. This reduction of the *STP* to the *NCP* demonstrates that the *NCP* is as hard to solve as the *STP*, meaning that this problem is NP-hard [8].

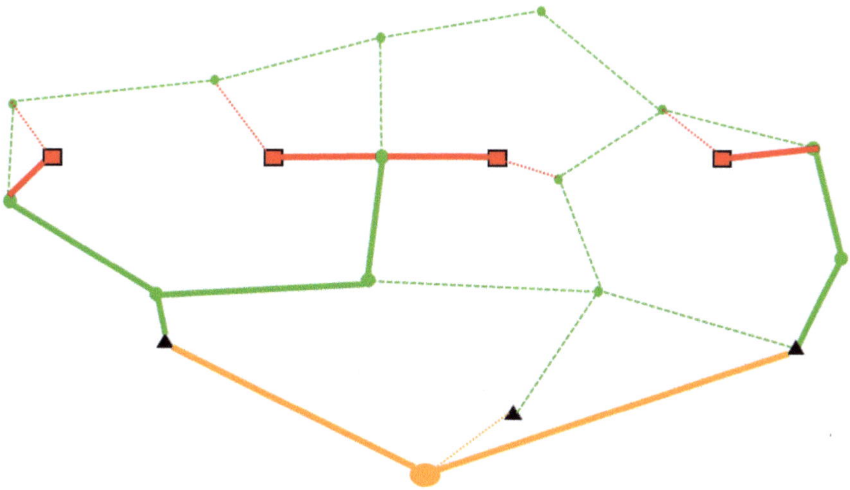

Fig. 3 The example of Figure 2 modeled as a *STP*. Nodes are represented in orange (s_0), as black triangles (*NR*), green circles (*NF*) and red squares (*S*); edges are represented in green (*EF*), yellow (*ER*) and red (*ES*). Bold edges represent the Steiner tree solution corresponding to the solution displayed in Figure 2

3 Case Study

The *Vale do Sousa* forest is located in the north of Portugal, in an area with a high population density, and covers a planted area of 7,623 hectares, primarily with eucalyptus and maritime pine. Over the years, this forest has experienced several fires. It comprises 655 management units owned by 185 proprietors, who are organized into an association that jointly manages the area, known as the *Paiva Forest Intervention Zone* (ZIF Paiva) [1].

Being a high population density area, the forest is crossed by a network of communal roads. Since it consists of smallholdings, new forest roads should preferably be constructed along the borders of the management units [5]. Therefore, the edges in *EF* and the nodes in *NF* are defined based on the boundaries of the management units, while the nodes in *NR* correspond to their intersections with existing communal roads (Fig. 4).

In this study, to determine the areas requiring protection, the forest area was divided into a grid of 1-hectare squares. A square is considered "protected" if the (Euclidian) distance from its geographical center to the nearest road does not exceed 200 meters (value suggested in [25]). This criterion identified 9117 sites to be accessed (i.e., the squares containing forest).

Several of these sites are within 200 meters of an existing road and can therefore be excluded from the problem, as they are already accessible. There are a few sites (40) in larger management units that are more than 200 meters away from any potential forest road. To access these sites, inward roads have to be built within management

Fig. 4 Boundaries of management units defining the potential forest roads in *Paiva Forest Intervention Zone* (ZIF Paiva)

units. These roads represent a loss to the owner of the land, so their cost is in general higher than the one of the forest roads bordering management units. For each site $i \in S$ we have selected a set of up to eight nodes $a \in \text{NF}$ that are at a distance of at most 400 meters from i. The new segments $\{a, i\}$ are added to ES.

After removing the sites that are already accessible via existing roads, the resulting problem comprises 2878 sites in S, 31157 edges in EF, and 29708 nodes in NF, from which 1622 are in NR.

The determination of costs of building roads is based in documents CAOF (Custos de Operações de Arborização, Rearborização e Beneficiação de Povoamentos Florestais) published in 2022 by the Portuguese Forest and Conservation Institute [11].

The cost of building a road segment $\{a, b\}$ depends on its length and also on the slope, which is computed as in [5], taking into account the slope of the line segment between nodes a and b. Based on published costs, 3 classes of slope have been considered, corresponding to 3 values for unit costs. For slopes (in percentage) in the intervals [0, 5], [5, 25] and [25, 45] the unit cost per Kilometer is respectively 1850, 4163.2 and 6476.4 Euro. Then the cost of building $\{a, b\}$ is obtained by multiplying its length in Kilometers by the corresponding unit cost.

To compute an estimate of the accessibility cost to a site $i \in S$, from a node $a \in \text{EN}$ located more than 200 meters away, the slope of the segment $\{a, i\}$ is used to determine the slope class, and the corresponding unit construction cost is doubled in order to penalise road construction within a management unit. The accessibility cost is then calculated as this new unit cost multiplied by the portion of the distance between a and i that exceeds 200 meters. These costs are considered only for sites located more than 200 meters from the nearest potential road.

Given the large scale of the problem to solve, we perform a geographical decomposition of the forest.

As shown in Fig. 5, the municipal roads divide the forest into several sub-regions, and the sites in S within each sub-region can only be accessed by roads constructed within that sub-region. Consequently, 37 sub-regions were identified, such that the union of the optimal solutions to the respective sub-problems constitutes the optimal solution to the overall case study problem. The number of sites to be accessed in these sub-regions varies from 1 to 900 (Table 1).

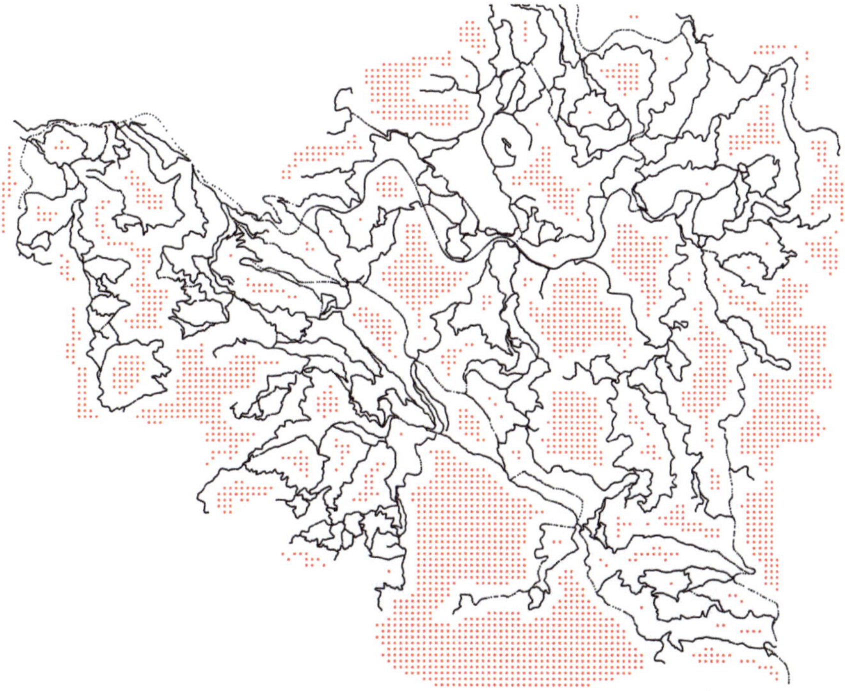

Fig. 5 ZIF Paiva: the existing communal roads (black) and the sites (red) that are farther than 200 meters from these roads. The roads define sub-regions that can be treated independently of each other

Table 1 Characteristics and solutions of the *NCP* instances defined by sub-regions in Paiva forest. For each sub-region it is displayed: the number of sites (#sites); the number of road segments (#*NCP* edges); the number of edges (#*STP* edges) and nodes (#*STP* nodes) in the corresponding *STP* formulation; the time in seconds for SCIP-Jack to solve the *STP* instance (*STP* time); the cost, in Euro, of the optimal solution (*STP* cost)

Subregion	#sites	#*NCP* edges	#*STP* edges	#*STP* nodes	*STP* time (secs)	*STP* cost (Euro)
A	19	601	1813	603	1	4441.68
AA	1	202	265	198	<1	114.33
AB	79	789	3118	848	<1	16333.63
AC	55	769	2310	788	< 1	15773.56
AD	339	2634	14544	2891	40	78499.30
	342	2637	14568	2894	36	79568.22
AE	104	1835	6319	1936	8	24474.07
	105	1836	6327	1937	7	24781.06
AF	16	895	2164	905	<1	6034.11
AG	75	504	2748	557	<1	16367.08
AH	34	762	1356	771	<1	11259.65
AI	26	1122	1783	1094	<1	9648.43
AJ	23	253	543	263	1	9435.09
AK	255	2172	10033	2354	9	49185.96
AK1	153	1160	5469	1274	4	26972.41
AK2	102	1100	4660 ara>	1172	4	22203.24
AL	2	53	68	56	1	816.98
AM	900	2818	19033	3575	19	176071.75
	913	2831	19137	3588	21	182316.32
B	4	257	454	249	<1	912.71
C	23	468	1215	469	<1	5418.67
D	16	982	1907	976	<1	3632.83
E	246	3496	13317	3602	44	49413.32
E1	68	2003	5566	1989	1	15460.37
E2	178	1536	7799	1657	15	33952.95
F	35	580	2523	596	1	9683.46
G	22	305	862	330	<1	5055.21
	31	314	934	339	<1	10542.76
H	19	771	1421	752	<1	8123.83
I	29	562	1543	573	1	8221.56
J	11	294	716	291	<1	2350.73
k	12	515	1412	517	<1	2119.46
L	9	323	685	323	1	1550.31

(continued)

Table 1 (continued)

Subregion	#sites	*#NCP* edges	*#STP* edges	*#STP* nodes	*STP* time (secs)	*STP* cost (Euro)
M	16	315	822	328	<1	2870.02
N	15	291	447	305	<1	2833.86
	19	295	479	309	<1	4615.75
O	33	642	1757	668	<1	11098.12
P	140	1347	6126	1444	5	31153.35
Q	17	483	824	500	<1	4532.02
R	30	554	1132	558	1	5371.53
S	99	1371	5040	1431	<1	20003.64
	109	1381	5120	1441	<1	27010.34
T	5	207	544	210	<1	1869.20
U	34	1390	2727	1373	1	7699.86
V	4	270	377	273	1	1218.31
XY	34	1937	3556	1925	<1	8305.09
Z	59	766	3225	790	<1	12123.49

4 Results

To solve the problem instances, we have modeled them as Steiner Tree Problems as described in section 2. The software package we used to solve the *STP* is called *SCIP-Jack* which has been developed by Daniel Rehfeldt and co-authors [19, 20]. *SCIP-Jack* addresses the Steiner tree problem by first applying specialized preprocessing techniques to simplify the problem instance, such as removing redundant nodes and edges. Then it transforms the problem into a directed Steiner arborescence problem and formulates it as a Mixed-Integer Linear Program. It then utilizes a sophisticated branch-and-cut algorithm (based on SCIP's internal Linear Programming solver, SoPlex [9]) to find optimal solutions. *SCIP-Jack* has demonstrated superior performance in solving the Steiner Tree Problem and variants, outperforming general purpose state-of-the-art Mixed Integer Programming solvers [18].

The problem instances we consider were obtained from decomposition of the Paiva forest as described in the previous section. The forest decomposes into 37 subregions, denoted by A,..., W, XY, Z, AA,..., AM. By decomposing further regions E and AK, four more instances were obtained, E1, E2, AK1 and AK2. Six of the instances possess sites which are farther than 200 meters from the nearest node in *NF*: G, N, S, AD, AE and AM. For these instances we consider two runs, one where those sites were discarded from the problem and another where they were included and could be accessed at an extra cost. In Table 1 the main characteristics of each instance are displayed. The columns *#sites* and *#NCP edges*, represent, respectively, the number of sites (#*S*) and number of forest edges (#*EF*).

Fig. 6 *NCP* AK1 instance. Sites in *S* are shown in red, the communal existing roads in orange, the potential forest roads in green and the nodes in *ER* as black triangles

For illustrative purposes we show instance AK1 in Fig. 6, displaying the sites in *S* (red), the communal existing roads (orange), the potential forest roads (green) and the nodes in *ER* (black triangles).

Figure 7 displays the *STP* instance corresponding to AK1. Edges in *EF* and nodes in $NF \backslash NR$ are displayed in green, nodes in *NR* are displayed as black triangles, nodes in *S* and edges in *ES* are displayed in red. Extra node s_0 and edges in *ER* connecting s_0 to nodes in *ER* are omitted in the figure.

To perform the tests, an 11th Gen Intel(R) Core(TM) i7-1165G7 @ 2.80GHz 2.80 GHz (16.0 GB RAM) was used.

Results for all the instances are also shown in Table 1. For instances G, N, S, AD, AE and AM, the first and second lines correspond respectively to the run where sites farther than 200 meters from the nearest node in *NF* are discarded and included.

All the instances have been solved in less than 44 seconds. In general, the cost of the solutions and the time required to obtain them increase with the number of sites and the number of edges in the *NCP* instances. The solution to the complete *NCP* problem for the Paiva forest can be obtained by merging the solutions of the 37 sub-regions. The total cost of this solution is 952850 Euro and the total time to obtain it (sum of the times required to solve the partial problems) is only 212 seconds.

Figure 8 displays the optimal solution for instance AK1. Built road segments are displayed in bold green. Edge segments that were not built are represented in dashed green. Sites in *S* are displayed in red and edges connecting them to the nearest node

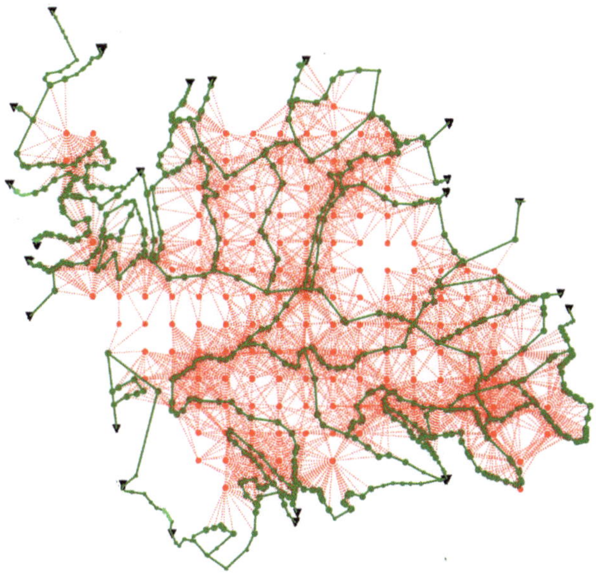

Fig. 7 *STP* graph for AK1. Edges in *EF* and nodes in $NF \backslash NR$ are displayed in green, nodes in *NR* are displayed as black triangles, nodes in *S* and edges in *ES* are displayed in red. Node S_0 and edges in *ER* are not shown

of a built segment are also represented in red. As before, communal existing roads are displayed in orange and the nodes in *ER* are black triangles.

In Figs. 9 and 10 a (small) sector of instance AM is displayed, showing the differences in solutions for the two runs of that instance (without and with sites farther than 200 meters from the nearest node in *NR*).

Instance AM has the largest number of sites located farther than 200 meters from the nearest forest node. We observe some differences between the solutions of the two runs. In general, for all cases with these extra sites, the resolution times for the two runs are similar.

5 Concluding Remarks

This work proposes modeling the *NCP* (Network Covering Problem) as an *STP* (Steiner Tree Problem) and solving the instances using a specialized state-of-the-art algorithm.

The main contributions of this article can be summarized as follows: (1) the modeling of the access road network design problem for firefighting (or the Network Covering Problem) as a Steiner Tree Problem, and (2) the demonstration of its resolution using a state-of-the-art algorithm (SCIP-Jack) applied to a real-world case study.

Fig. 8 Optimal solution for AK1. Built road segments are displayed in bold green. Edge segments that were not built are represented in dashed green. Sites in S are displayed in red as well as edges connecting them to the nearest node of a built segment. As before, communal existing roads are displayed in orange and the nodes in ER are black triangles

Computational results demonstrate the success of the proposed approach. All problems in a case study of a forest in northern Portugal were solved in less than one minute on a laptop. We believe that this approach could also be effective for larger or more complex problems. For instance, it could handle scenarios where roads are not strictly confined to the boundaries of management units, allowing for greater flexibility in layout design; where the network of pre-existing roads is smaller; or where a refined, narrower grid of sites is used to ensure more uniform accessibility to the forest.

This exploratory work provides a foundation for several future extensions.

In the case study, the selection of points to protect or access was based on a fixed distance of 200 meters. However, it is possible to adapt this distance according to terrain characteristics. For example, the distance may be shorter if the site to be accessed is at a higher elevation than the road. Additionally, other criteria could be incorporated, such as weighting the distance based on site value or fire risk. Sites with higher value or greater fire risk might justify shorter distances to roads.

An important factor in designing access roads for forest firefighting is the access time to protected sites. No limit on the length of roads was imposed, and some of the solutions obtained contain long roads to access some of the sites to protect. A potential extension of this work could involve incorporating minimum access time constraints into the model.

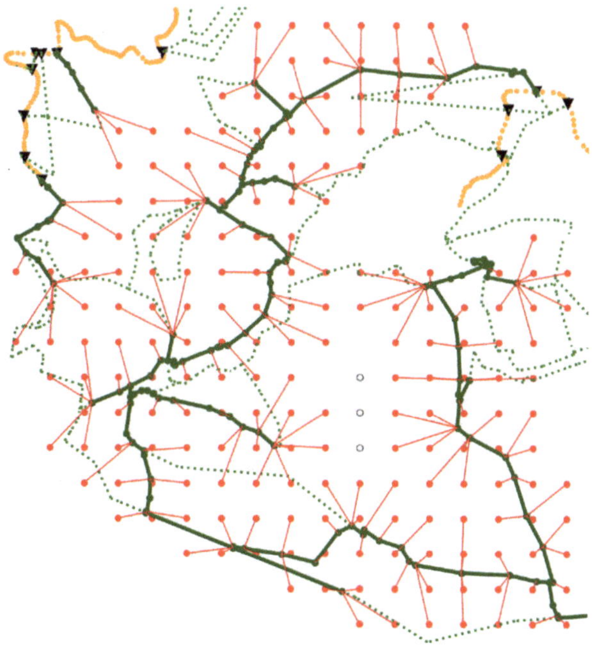

Fig. 9 Detail of AM solution for the run without sites farther than 200 meters from the nearest road node (three sites displayed as black circumferences)

In this problem, as in most forest road construction problems, the optimal network for accessing multiple forest points tends to form a tree structure (in graph theory terms), resulting in a set of dead-end paths from existing roads. However, to ensure safe wildfire suppression, roads should ideally have at least two exits, allowing vehicles to move freely and increasing the chances of escape in emergencies [12]. An extension of this work could involve designing road networks that meet this critical safety requirement.

One of the primary uses of forest roads is to provide access to management units for timber extraction and other silvicultural activities [5]. Another natural extension of this research would involve developing models that account for road construction considering both logging and silvicultural needs, and also ensuring access for firefighting.

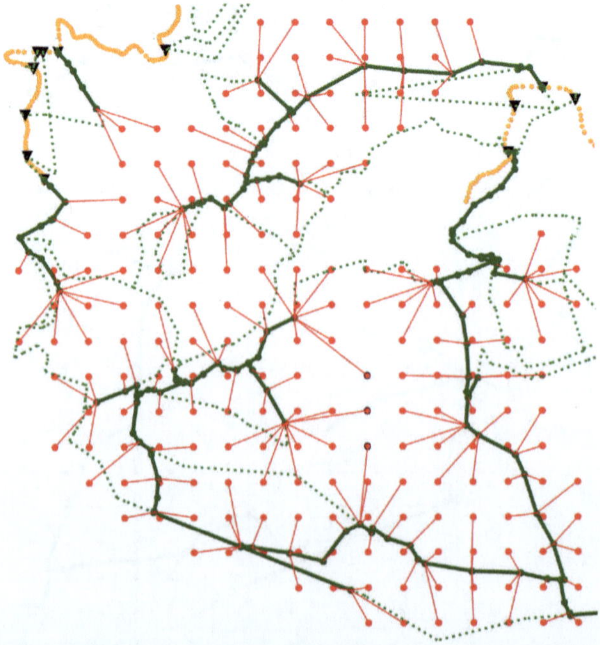

Fig. 10 Detail of AM solution for the run with sites farther than 200 meters from the nearest road node

Acknowledgements We thank Susete Marques from Instituto Superior de Agronomia, Universidade de Lisboa, for providing the GIS data used in this study. This study was partially funded by the Center for Mathematical Studies, University of Lisbon (CEMS.UL), under the project reference UID/04561/2025, financed by the Fundação para a Ciência e a Tecnologia, I.P. (FCT).

References

1. Associação Florestal do Vale do Sousa (ZIF-Paiva): https://www.afvs.ws/zif-paiva.html (1994)
2. Akay, A. E., Kara, I. R., Kahraman, I.: Determining the locations of potential firefighting teams by using GIS techniques. Int. Arch. Photogramm. Remote Sens. Spat. Inf. Sci. - ISPRS Arch. 42.4/W9, 83–88 (2018) https://doi.org/10.5194/isprs-archives-XLII-4-W9-83-2018
3. Akay, A. E., Wing, M. G., Sivrikaya, F., Sakar, D.: A GIS-based decision support system for determining the shortest and safest route to forest fires: A case study in Mediterranean Region of Turkey. Environ. Monit. Assess. **184**(3), 1391–1407 (2012) https://doi.org/10.1007/s10661-011-2049-z
4. Campbell, M. J., Page, W. G., Dennison, P. E., Butler, B. W.: Escape route index: A spatially-explicit measure of wildland firefighter egress capacity, Fire **2**(3), 1–19 (2019) https://doi.org/10.3390/fire2030040
5. Constantino, M.F., Mesquita, M., Marques, S., Tóth, S.F., Borges, J.G.: Road network, landing location, and routing optimization for forest smallholders landscapes, Intl. Trans. in Op. Res. **32**, 888–917 (2025). https://doi.org/10.1111/itor.13485

6. Diestel, R.: Graph Theory, Graduate Texts in Mathematics, Volume 173, Springer-Verlag (2025) https://doi.org/10.1007/978-3-662-70107-2
7. Dimopoulou, M. and Giannikos, I.: Towards an integrated framework for forest fire control, Eur. J. Oper. Res. 152.2, 476–486 (2004) https://doi.org/10.1016/S0377-2217(03)00038-9
8. Garey, M.R., Johnson, D.S.: Computers and intractability: A guide to the theory of NP-completeness. W. H. Freeman and Company (1979). ISBN: 978-0716710455
9. Gleixner, A.M., Steffy, D.E., Wolter, K.: Iterative refinement for linear programming. INFORMS J. Comp. **24**(3), 470–483 (2012). https://doi.org/10.1287/ijoc.1110.0463
10. Hwang, F.K., Richards, D.S., Winter, P.: The Steiner Tree Problem, North-Holland (1992). ISBN: 978-0-444-88045-6
11. ICNF: CAOF—Custos de Operações de Arborização, Rearborização e Beneficiação de Povoamentos Florestais, Tabelas CAOF, https://www.dgadr.gov.pt/mecanizacao/caof (2022)
12. Laschi, A., Foderi, C., Fabiano, F., Neri, F., Cambi, M., Mariotti, B., Marchi, E.: Forest road planning, construction and maintenance to improve forest fire fighting: A review. Croat. J. For. Eng. **40**(1), 207—219 (2019). https://crojfe.com/archive/volume-40-no-1
13. Liampas, S.A.G., Stamatiou, C.C., Drosos, V.C.: Forest road network planning for biomass exploitation and fire preventions: A least cost path analysis, Agric. Eng. Int. CIGR J. **21**(4), 33–42 (2019). https://cigrjournal.org/index.php/Ejounral/article/view/5385
14. Lourenço, M., Oliveira, L.B., Oliveira, J. P., Mora, A., Oliveira H., Santos, R.: Integrated decision support system for improving wildfire suppression management, ISPRS Int. J. Geo-Inf. **10**(8) (2021). https://doi.org/10.3390/ijgi10080497
15. Murray, A.T.: Route planning for harvest site access, Can. J. For. Res. **28**(7), 1084–1087 (1998) https://doi.org/10.1139/x98-12
16. Podolskaia, E., Ershov, D., Kovganko, K.: Automated construction of ground access routes for the management of regional forest fires, J. For. Sci. **66**(8), 329–338 (2020). https://doi.org/10.17221/59/2020-JFS
17. Psilovikos, T.A., Doukas, K.G., Drosos, V.K.: The contribution of forest roads to the forest fire protection, Formec 2011 - Push. boundaries with Res. Innov. For. Eng. Dimitrakopoulos 2000, 1–13 (2011). https://www.formec.org/images/proceedings/2011formec2011_paper_psilovikos_etal.pdf
18. Rehfeldt, D.,Koch, T.: SCIP-Jack–A solver for STP and variants: Mathematical programming computation. Math. Program. Comp. **11**(2), 302–329 (2019). https://doi.org/10.1007/s12532-018-0143-1
19. Rehfeldt, D., Shinano, Y,. Koch, T.: SCIP-Jack: An Exact High Performance Solver for Steiner Tree Problems in Graphs and Related Problems, in Modeling, Simulation and Optimization of Complex Processes HPSC 2018, 201–223 (2021) https://doi.org/10.1007/978-3-030-55240-4_10
20. Rehfeldt, D., Koch, T.: Implications, conflicts, and reductions for Steiner Trees. Math. Program. **197**, 903–996 (2023). https://doi.org/10.1007/s10107-021-01757-5
21. Stefanovi, B., Stojni, D., Danilovi, M.: Multi-criteria forest road network planning in fire-prone environment: a case study in Serbia. J. Environ. Plan. Manag. **59**(5), 911—926 (2016). https://doi.org/10.1080/09640568.2015.1045971
22. Sullivan, P.R., Campbell, M.J., Dennison, P.E., Brewer, S.C., Butler, B.W.: Modeling wildland firefighter travel rates by terrain slope: Results from GPS-tracking of type 1 crew movement. Fire **3**(3), 1–14 (2020). https://doi.org/10.3390/fire3030052
23. Ta, Í, Akay, A.: Examining the Road Networks and Locations of Firefighting Teams by Using GIS Techniques, in Proceedings of the Forestry Academy of Sciences of Ukraine, vol 21, 173–180 (2020) https://doi.org/10.15421/412037
24. Thompson, M.P., Gannon, B.M., Caggiano, M.D.: Forest roads and operational wildfire response planning. Forests **12**(2), 1–11 (2021). https://doi.org/10.3390/f12020110
25. Zhang, F., Dong, Y., Xu, S., Yang, X., Lin, H.: An approach for improving firefighting ability of forest road network. Scand. J. For. Res. **35**(8), 547–561 (2020). https://doi.org/10.1080/02827581.2020.1829029

26. Zong, X., Tian, X., Wang, X.: An optimal firebreak design for the boreal forest of China. Sci. Total Environ. **781**, 146822 (2021). https://doi.org/10.1016/j.scitotenv.2021.146822

The Graph Burning Problem under Constrained Diffusion

Enrico Iurlano, Günther R. Raidl, and Marko Djukanović

Abstract Relying on a simplified model of fire spread, the Graph Burning Problem is an NP-hard combinatorial optimization problem that yields a metric of social contagion or vulnerability of a network. It concerns a discrete-time process on a simple undirected graph, with, in each timestep, a diffusion phase of fire towards all neighbors of "burned" vertices and a second phase, in which a next not-yet-burned vertex is ignited by an external actor. The aim is to find the minimum number of timesteps together with suitable choices of vertices for the actor such that the whole set of vertices gets burned. The applicability of the problem becomes more relevant when one takes into account that diffusion might realistically be suppressed by given natural obstacles and limitations as well as implemented countermeasures. Therefore, this paper proposes the Constrained Diffusion Graph Burning Problem, where vertex-specific thresholds are considered that determine the maximum number of neighbors the vertex can ignite. We consider the aspect of expiration, i.e., burned vertices are permitted to diffuse fire only immediately after becoming burned, but not to a later timestep. These assumptions not only appear meaningful in the context of fire spread but also within dynamics of viral contagion and rapid information dissemination. A time-indexed (mixed) integer linear programming approach is proposed in two versions: The first version one-hot encodes the time, and the second one handles the time using a Miller-Tucker-Zemlin formulation. Finally, a greedy heuristic is proposed and compared to the previous formulations.

Keywords Social contagion model · Fire spread simulation · Mixed integer linear programming

---

E. Iurlano (✉) · Günther R. Raidl
Algorithms and Complexity Group, TU Wien, Vienna 1040, Austria
e-mail: eiurlano@ac.tuwien.ac.at

Günther R. Raidl
e-mail: raidl@ac.tuwien.ac.at

M. Djukanović
Faculty of Natural Sciences and Mathematics, University of Banja Luka, Banja Luka 78000, Bosnia and Herzegovina
e-mail: marko.djukanovic@pmf.unibl.org

F. Alvelos et al. (eds.), *Advances in Optimization and Wildfire*, Lecture Notes in Operations Research, https://doi.org/10.1007/978-3-032-03108-2_9

1 Introduction

In 2014, Bonato et al. [1] proposed the discrete-time process of *graph burning* as a model of social contagion. This process relies on a highly simplified model of fire spread and originally was used to provide insights into the achievability of rapid information dissemination in social networks under certain temporal assumptions on the release of information. Due to similarities, e.g., to the *Target Set Selection Problem* [2] or the *Least Cost Influence Maximization Problem* (LCIMP) [3], the *Graph Burning Problem* (GBP) additionally provides an interesting alternative model for viral marketing or opinion-making. For a formal description of the graph burning process, we consider a simple and undirected input graph $G = (V, E)$ whose vertices $v \in V$ carry a time-dependent binary status indicating either *unburned* or *burned* [1]. Assuming (discrete) timesteps indexed by $t \in \mathbb{N} \cup \{0\}$, consider the following process.

- Initially, all vertices are unburned.
- In timestep $t = 0$ the status of a single vertex is changed to burned.
- In each timestep $t \geq 1$, the neighbors of all vertices that have the status burned in timestep $t-1$ become burned as well. Note that some of them may already be burned.
- Within the same timestep t, an unburned vertex is picked and made burned (if such a one exists).

The process is iterated and stops when a timestep is reached after which all the vertices of the graph are burned. Each such process is called an admissible burning process. The *burning number* $b(G)$ of G is defined as the minimum number of timesteps over all admissible burning processes, i.e., $b(G) := \min\{T+1 :$ there is an admissible burning process on G terminating with index $T\}$. Returning to the perspective of a social contagion model, a burned vertex could represent—within the setting of opinion-making or viral marketing—the scenario in which an individual has been convinced to adopt a certain opinion either by a direct influence of some external actor or by propagation from a neighboring individual that is already affected by that opinion. A small burning number for a social network reveals that it is possible to influence all individuals in a short amount of time.

Several combinatorial aspects of the graph burning number have been examined, e.g., it is known that its calculation is NP-hard even on trees of maximum degree three [4] or on caterpillars of maximum degree three [5, 6]; several other graph classes are mentioned in the literature in this context, see [7] for a partial overview. The central open problem in the field is stated in [1] claiming that for any graph $G = (V, E)$ the bound $b(G) \leq \lceil |V|^{1/2} \rceil$ holds; partial progress towards this so-called *burning number conjecture* has been recently obtained; in [8] it is proved that this conjecture holds asymptotically, and in [9] it is shown that the conjecture holds for any tree without vertices of degree two. The survey of Bonato [7] describes several further interesting computational findings on $b(G)$, in particular, established $b(G)$-numbers respectively approximation algorithms [10]; for general graphs, a 3-approximation algorithm is presented together with a 2-approximation algorithm specifically designed for trees. A randomized 2.314-approximation algorithm for computing the burning number of a general graph is presented in [11]. In [12], an approximation algorithm termed "burning farthest-first" is proposed with a tighter approximation ratio of $3 - 2/b(G)$. It relies on the idea to iteratively incorporate locally optimal vertices into an initially empty set of candidate vertices. The chosen vertex lies here among the vertices that maximize the distance to the set of burning vertices. A similar approach has been designed earlier in [13] but from a purely heuristic perspective. Several further heuristics for this problem are proposed in [14, 15] based on the eigencentrality as a measure for attractiveness of a next vertex to be ignited. More recently in [16], based on the relationship between the GBP and the so-called *Clustered Maximum Coverage Problem*, the authors propose a greedy heuristic that effectively solves instances with up to hundreds of thousands of vertices. An integer linear programming formulation and a constraint programming approach are proposed in [17]. Recently, in [18] considerable progress has been achieved towards solving large solution instances optimally by a row generation approach.

The main contribution of this paper is the introduction and solution of a generalization of GBP where vertices can ignite a certain maximum number of neighbors. The generalization enlarges the search space of the original problem and requires the representation of feasible solutions to carry additional information. We propose two Mixed Integer Linear Programming (MILP) formulations and evaluate their computational advantages and limitations by running experiments on instances from the literature commonly used for the classical GBP. Furthermore, we propose a greedy approach which can be executed in one mode by solving subproblems via these formulations or, alternatively, by using a degree- and eigencentrality-based scoring function.

Notation and preliminaries. If not stated otherwise, simple and undirected graphs $G = (V, E)$ are considered; the *order* of a graph G refers to $|V|$, the number of vertices. By $N(v)$, for $v \in V$, we denote the set of vertices that are adjacent to v, i.e., $N(v) := \{w \in V : \{v, w\} \in E\}$. Moreover, let us denote by $\hat{G}$ the digraph resulting from bidirecting each edge of G, i.e., $\hat{G} := (V, \hat{E})$ with $\hat{E} := \{(u, v), (v, u) \in V \times V : \{u, v\} \in E\}$. In case some graph $G = (V, E)$ is a directed graph (digraph), the *in-degree* and *out-degree* of v refer to the quantities $|\{u \in V : (u, v) \in E\}|$ and $|\{w \in V : (v, w) \in E\}|$ respectively. The number of edges on a shortest path between v and w is denoted by $\text{dist}(v, w)$. An *arborescence* is a directed acyclic graph $A = (V, E)$ with at most one vertex of in-degree zero—called the *root*—having in-degree one for all remaining vertices. Denote by leaves(A) the set of all leaves, i.e., vertices with out-degree zero, by $\text{depth}(v)$ the *depth*, i.e., the number of edges on the unique directed path from the root of A to a vertex v, and by $\text{height}(A)$ the maximum depth over all vertices contained in A. Sequences (or unions) of pairwise vertex-disjoint arborescences are called *forests of arborescences*. Denote by $\lfloor \cdot \rfloor$ and $\lceil \cdot \rceil$ the *floor* and *ceil* function, respectively.

The remainder of the paper is organized as follows. In Sect. 2 we motivate our generalization of graph burning. Section 3 provides two MILP formulations to solve the introduced problem. Section 4 proposes a heuristic approach whose performance is in Sect. 5 compared to the MILP approaches. Conclusions and open questions are given in Sect. 6.

2 Constraints on the Diffusion Process

The classical GBP despite being practically motivated, assumes for its diffusion process that all neighbors of burned vertices get burned in the subsequent timestep. For many scenarios this assumption may seem too unrealistic: In general, people can successfully influence only a certain maximum number of other people from their circle of contacts (e.g., due to different individual resources, skills, or ambition for persuasion). Therefore, two constraints on the diffusion process shall be imposed to make the model more realistic and more flexible. We propose to limit, by a vertex-dependent threshold, the number of neighbors that can be ignited by a burned vertex.

Moreover, we want to incorporate the assumption that people become active in influencing other people immediately after they have been influenced; we motivate this by referring to [19] which demonstrates that the spread of social contagion is reduced by the decay of novelty. In the language of fire diffusion, the latter model is explained as follows: Vertices represent regions and a comparable behavior of the fire spread arises when fire can be contained to spread to a maximum number of neighboring regions, e.g., by having a specific local resource of firefighter units. Moreover, in this context, we assume that an already burned region cannot propagate fire at a later time point.

For a formalization, let us consider the graph $G = (V, E)$ with imposed thresholds $\theta_v \in \mathbb{N} \cup \{0\}$, $v \in V$; these thresholds are collected in a vector of parameters $\theta = (\theta_v)_{v \in V}$. Apart from the aforementioned time-dependent status burned or unburned for each vertex, we implicitly track a second time-dependent, irreversible binary status of a vertex, called the *expiration* status, which prevents vertices that have this status from transmitting their burned status to their neighbors. Assume that we are given a maximum time index $T \in \mathbb{N} \cup \{0\}$. Initially, for $t = 0$, burn a single vertex. Afterwards, execute the following actions for $t = 1, \ldots, T$:

(i) *Constrained diffusion from unexpired vertices:* For each vertex v that has the status burned in timestep $t - 1$ but not in timestep $t - 2$, burn at most θ_v of its—according to the timestep $t - 1$—unburned neighbors. (By convention, in the (-1)-th timestep all vertices are unburned.)
(ii) *Burn:* Pick at most one currently unburned vertex and make it burned.

We call an execution of the latter procedure a *T-terminating θ-burning process (with constrained diffusion)* and denote the set of all such processes as $\mathcal{B}_{\theta,T}(G)$. For $P \in \mathcal{B}_{\theta,T}(G)$, the number of vertices with the status burned after termination of both phases in the final timestep T is called the *penetration* of P, adopting the terminology of the LCIMP [3]. If T^* is the minimum number that ensures a process $P \in \mathcal{B}_{\theta,T^*}(G)$ of penetration $|V|$ exists, we define $b_\theta(G) := T^* + 1$ and call the latter *(constrained diffusion) θ-burning number of G*. The problem of determining this optimal value T^*

is henceforth called the *Constraint Diffusion Graph Burning Problem* (CDGBP, or θ-GBP). If we want to provide more information on the burning process, we denote it as $\mathcal{B}_{\theta,T}(G, (s_0, \ldots, s_T), y)$ where s_t is the chosen vertex, henceforth called *seed*[1] during the *burn*-phase in timestep t. Moreover, for each directed edge $(u, v) \in \hat{E}$ we define

$$y_{u,v} := \begin{cases} 1 \text{ if } u \text{ ignited } v \text{ in some timestep } t \in \{1, \ldots, T\}, \\ 0 \text{ otherwise.} \end{cases}$$

The subsequent Proposition 1, in particular, gives a preparatory viewpoint for the MILP formulations stated later by directly lifting the considerations concerning $b(\cdot)$ in the proof of [1, Theorem 5] to θ-burning.

Proposition 1 Let $T \in \mathbb{N} \cup \{0\}$ and $X \subseteq V$ for a given graph $G = (V, E)$ with thresholds θ. Then, the following assertions are equivalent.

(i) There is a θ-burning process $\mathcal{B}_{\theta,T}(G, (s_0, \ldots, s_T), y)$ whose set of burned vertices after termination is X.
(ii) In G there is a subforest consisting of arborescences $T_{s_0}, T_{s_1}, \ldots, T_{s_T}$ with respective roots $s_0, \ldots, s_T$ and heights $h_0, \ldots, h_T$ such that, firstly, the set of all vertices contained in some arborescence corresponds to X; secondly, $h_t \leq T - t$ for $t = 0, \ldots, T$; and thirdly, each $v \in X$ is a member of a single arborescence $T_{s(v)}$ where the out-degree of v inside $T_{s(v)}$ is at most θ_v.

Proof (i) $\Longrightarrow$ (ii): Let $P \in \mathcal{B}_{\theta,T}(G, (s_0, \ldots, s_T), y)$ and X be the set of vertices burned after termination of P. During the process, a vertex v might have become burned because one vertex (or even multiple vertices) "chose" to make their neighbor v burned. In the following, consider the digraph $\hat{G} := (V, \hat{E})$, i.e., the bidirected version of G. Note that if $s \in \{s_0, \ldots, s_T\}$ is a seed, we have $y_{u,s} = 0$ for each $u \in N(s)$, whereas $y_{s,u}$ can in fact attain the value 1 for $u \in N(s)$. If $\hat{E}|_y := \{(u, v) \in \hat{E} : y_{u,v} = 1\}$, then $\hat{G}|_y := (V, \hat{E}|_y)$ contains no directed cycles (not even of length two): In fact, the existence of a vertex having an incoming and an outgoing edge contradicts the property that only unburned vertices can be turned to burned by a burned neighbor. Facing a directed acyclic graph, we can therefore locate a set of sources (i.e., vertices of in-degree zero) that here must coincide with $\{s_0, \ldots, s_T\}$.

We run now the following procedure which returns an updated edge-status $y'_{u,v} \in \{0, 1\}$ satisfying for all $(u, v) \in \hat{E}$ that $y'_{u,v} \leq y_{u,v}$: Initially set $y'_{u,v} := y_{u,v}$. Then, iteratively, for $d = 1, \ldots, T$, execute the following: For each vertex $v \in V$ being reachable from a seed s_t via a directed length-d path do the following: Pick the seed s_{t^*} with the minimum index t^* permitting such a directed path $(s_{t^*}, u_1, u_2, \ldots, u_{d-1}, v)$. Then, overwrite $y'_{u,v} := 0$ for all $u \in N(v) \setminus \{u_{d-1}\}$. After termination of this procedure, we obtain that $\hat{G}|_{y'} := (V, \hat{E}|_{y'})$ is a forest of arborescences $(A_{s_0}, \ldots, A_{s_T})$ whose roots are given by the seeds of the process. As all finally-burned vertices in

[1] Potentially a seed reflects the empty vertex selection ε and this will manifest itself in our models in that no activation from an external source vertex is present.

X got burned in timestep T the latest, the distance from s_t to any vertex in A_{s_t} can be at most $T - t$. Moreover, y meets the θ-thresholds for each vertex $v \in V$, i.e., $\sum_{w \in N(v)} y_{v,w} \leq \theta_v$, and hence y', as a thinned-out version of y, does so as well.

(ii) $\Longrightarrow$ (i): An admissible θ-burning process can be reconstructed by simply choosing the sequence of seeds $(s_0, \ldots, s_T)$ and in timestep t, for each burned vertex v, choosing as next vertices to be burned those being the children of v when v is seen as a member of its associated arborescence. Here the threshold-respecting out-degrees of the vertices in the arborescence will automatically imply compatibility with the thresholds for an admissible θ-burning process. The number of timesteps needed for this burning process is bounded from above by T, the height-restriction on the arborescence rooted in s_0. By construction, all $v \in X$ are therefore burned at most after termination of timestep T.

Although equipped with further details, Fig. 1 in the later Sect. 3 is useful for visualizing the last characterization.

Observation 1 For $G = (V, E)$ and $\theta_v := \deg(v), v \in V$, we have $b_\theta(G) = b(G)$.

Given $G = (V, E)$, θ, and T, finding the maximum penetration over all $P \in \mathcal{B}_{\theta,T}(G)$ may clearly be seen as a special case of the problem asking, for a pre-specified *arbitrary* parameter vector of heights $H := (h_1, \ldots, h_Q)$, $Q \leq |V|$, to find a sequence of Q arborescences maintaining these heights, respecting the thresholds θ and yielding maximum penetration. For the 1-tuple $H = (h)$ we call it the problem of finding a *Maximum Coverage Arborescence with θ-bounded Degree and h-bounded Height*, abbreviated MCADH(G, θ, h).

3 Mixed Integer Linear Programming Formulations

In the following, given the maximum time-index T, we address the problem of finding a sequence of seeds $(s_0, \ldots, s_T)$ for a θ-burning process leading to maximum penetration in a given graph $G = (V, E)$ with threshold parameter θ. The idea is to make use of Proposition 1 and to find a collection of pairwise vertex-disjoint arborescences $A^0, \ldots, A^T$ of the graph meeting all height and threshold constraints and maximizing the number of vertices covered by the collection.

An Integer Linear Program (ILP) purely relying on binary variables can be derived in the fashion of a time-indexed approach as done, e.g., in [12, 20] for related problems. We present it using an equivalent arborescence-based perspective: First, an external source vertex p with out-going directed edges pointing to all vertices in V is added to the graph. Introduce for all $t \in \{0, \ldots, T\}$ and $v \in V$ a depth-tracker $z_v^t \in \{0, 1\}$ indicating whether the arborescence covering vertex v, say its index is $s \in \{0, \ldots, T\}$, fulfills $s + \text{depth}(v) = t$. Similarly, for each $(v, w) \in \hat{E} \cup (\{p\} \times V)$ the variable $c_{v,w}^t \in \{0, 1\}$ indicates that $s + \text{depth}(v) = t - 1$ and, simultaneously, $s + \text{depth}(w) = t$ (using the convention $\text{depth}(p) = -1$). Here, the shifted depth-index $t = s + \text{depth}(v)$ agrees precisely with the timestep in which vertex v acquires the burned status (via diffusion or via a new ignition). Shifted depth-counters ensure that arborescences with a higher index are forced to attain lower heights. Based on these variables we derive the following formulation which we call "One-Hot encoded Depth" (OHD):

$$\max \sum_{t\in\{0,\dots,T\}} \sum_{v\in V} z_v^t \tag{1}$$

$$\text{s.t. } z_v^0 = c_{p,v}^0 \qquad \forall v \in V \tag{2}$$

$$\sum_{v\in V} c_{p,v}^t \leq 1 \qquad \forall t \in \{0, \dots, T\} \tag{3}$$

$$\sum_{t\in\{0,\dots,T\}} z_v^t \leq 1 \qquad \forall v \in V \tag{4}$$

$$\sum_{v\in N(w)} c_{v,w}^t \leq 1 \qquad \forall t \in \{0, \dots, T\}, \forall w \in V \tag{5}$$

$$\sum_{t\in\{0,\dots,T\}} c_{v,w}^t \leq 1 \qquad \forall (v, w) \in \hat{E} \tag{6}$$

$$\sum_{t\in\{0,\dots,T\}} \sum_{v\in V\cup\{p\}} c_{v,w}^t \leq 1 \qquad \forall w \in V \tag{7}$$

$$z_w^t \leq \sum_{v\in N(w)\cup\{p\}} c_{v,w}^t \qquad \forall t \in \{1, \dots, T\}, \forall w \in V \tag{8}$$

$$\sum_{w\in N(v)} c_{v,w}^t \leq \theta_v z_v^{t-1} \qquad \forall t \in \{1, \dots, T\}, \forall v \in V \tag{9}$$

$$c_{v,w}^t \leq z_v^{t-1} \qquad \forall t \in \{1, \dots, T\}, \forall (v, w) \in \hat{E} \tag{10}$$

$$c_{v,w}^t \in \{0, 1\} \qquad \forall t \in \{0, \dots, T\}, \forall (v, w) \in \hat{E} \cup (\{p\} \times V) \tag{11}$$

$$z_v^t \in \{0, 1\} \qquad \forall t \in \{0, \dots, T\}, \forall v \in V \tag{12}$$

Formulation (1)–(12) works as follows. The objective function (1) counts all vertices that are assigned a depth t and which thus must be part of one of the arborescences. Constraint (2), eliminable by preprocessing, associates depth 0 to a vertex v iff the 0-indexed activation from the source p to a vertex v is present. Constraint (3) ensures that for any activation index t, at most one vertex is activated by the source p. Constraint (4) forbids the assignment of more than one depth to each vertex. Constraint (5) guarantees that no matter which time index t is considered, the respective incoming activations to a vertex $w \in V$ from neighbors $v \in N(w)$ are at most one—uniqueness of predecessors in the arborescences is achieved in this way. The fact that an edge is active in at most one time step is represented by (6). Inequalities (7) enforce that no activations from potentially differing time indices point to a given vertex, as this is not forbidden solely by (5). The fact that for an index $t \geq 1$ and any vertex w no incoming t-indexed activation from any neighbor v (or the source p) is present implies that depth t is not assignable to w. To make sure that a vertex v does not activate more than θ_v neighbors, constraint (9) is added. Constraint (10) is a strengthening inequality affirming that activation at timestep t via an outgoing edge of v requires that v has been assigned a shifted depth of $t-1$.

Let us add a formulation by adapting the well-known Miller-Tucker-Zemlin (MTZ) approach [21]. The goal is here to employ only a single binary activation variable per directed edge instead of $T+1$ associated binary variables. We introduce $T+1$ external source vertices $p_0, \ldots, p_T$ all having out-going edges pointing to each vertex of the graph. For each $v \in V$ and $t \in \{0, \ldots, T\}$ we let $\ell_v^t \in \{0, 1\}$ indicate if v lies on A^t ($\ell_v^t = 1$ in the affirmative case, otherwise $\ell_v^t = 0$). Moreover, let the fractional variable $z_v \in [0, T]$ track the shifted depth of v with respect to its covering arborescence, and let $y_{v,w} \in \{0, 1\}$ indicate if vertex w is a successor of vertex v in their covering arborescence. We declare the following formulation as MTZ:

$$\max \sum_{t \in \{0,\ldots,T\}} \sum_{v \in V} \ell_v^t \tag{13}$$

$$\text{s.t.} \sum_{t \in \{0,\ldots,T\}} \ell_v^t \leq 1 \qquad \forall v \in V \tag{14}$$

$$\sum_{v \in V} y_{p_t,v} \leq 1 \qquad \forall t \in \{0, \ldots, T\} \tag{15}$$

$$y_{p_t,v} \leq \ell_v^t \qquad \forall t \in \{0, \ldots, T\}, \forall v \in V \tag{16}$$

$$y_{v,w} \leq \sum_{t \in \{0,\ldots,T\}} \ell_v^t \qquad \forall (v, w) \in \hat{E} \tag{17}$$

$$y_{v,w} \leq \sum_{t \in \{0,\ldots,T\}} \ell_w^t \qquad \forall (v, w) \in \hat{E} \tag{18}$$

$$\ell_w^t \leq y_{p_t,w} + \sum_{v \in N(w)} y_{v,w} \qquad \forall t \in \{0, \ldots, T\}, \forall w \in V \tag{19}$$

$$\ell_v^t - 1 + y_{v,w} \leq \ell_w^t \qquad \forall t \in \{0, \ldots, T\}, \forall (v, w) \in \hat{E} \tag{20}$$

$$z_v \geq t\ell_v^t \qquad \forall t \in \{0, \ldots, T\}, \forall v \in V \tag{21}$$

$$z_v + 1 \leq z_w + (1 - y_{v,w})(T + 1) \qquad \forall (v, w) \in \hat{E} \tag{22}$$

$$0 \leq z_v \leq T \qquad \forall v \in V \tag{23}$$

$$\sum_{w \in N(v)} y_{v,w} \leq \theta_v \qquad \forall v \in V \tag{24}$$

$$y_{v,w} \in \{0, 1\} \qquad \forall (v, w) \in \hat{E} \cup (\{p_0, \ldots, p_T\} \times V) \tag{25}$$

$$\ell_v^t \in \{0, 1\} \qquad \forall t \in \{0, \ldots, T\}, \forall v \in V \tag{26}$$

$$z_v \in \mathbb{Q} \qquad \forall v \in V \tag{27}$$

This formulation relies on a fractional counter variable that follows the rule of incrementation, see (22), up to a maximum height (23). If vertices are covered by the t-indexed arborescence, then their depths will be enumerated starting from t—this is imposed by (21). Constraint (24) implements the local bounds θ_v. The significant difference to the formulation OHD appears in (15)–(20), where the arborescence-indicating label resulting from ℓ must be inherited by activated neighbors; in fact, y is a variable shared by all arborescences, therefore propagating the right arborescence-labels is crucial. An illustration for MTZ is given in Fig. 1. The MTZ formulation can be strengthened by distance-based constraints: Vertices separated by a distance greater than the arborescence's height cannot both belong to the arborescence; a precalculation of the distances between all pairs of vertices is therefore required and can be achieved by Johnson's algorithm. Such an approach has been previously pursued in the setting of [17]. More precisely, we can impose

$$y_{p_t,v} + \ell_w^t \leq 1 \quad \forall t, \ \forall \{v, w\} \in \{\{v', w'\} \subseteq V : \mathrm{dist}(v', w') > T - t\}. \tag{28}$$

Remark 1 The proposed formulations rely on a fixed maximum time-index T. If we are interested in calculating b_θ, or more precisely in finding a shortest sequence of arborescences yielding full penetration, in principle, we can start with $T = 0$ and determine by a MILP solver if full penetration is already achievable. If this is not the case we iteratively repeat the procedure with incremented value of T, until a desired sequence of arborescences is found. Technically, we only need to solve a feasibility MILP where the original objective function is replaced with the constraint enforcing that all $|V|$ vertices must be burned. To reduce the number of MILP solver calls, it is conceivable to perform a binary search on the minimum value for T as in [17, Algorithm 1].

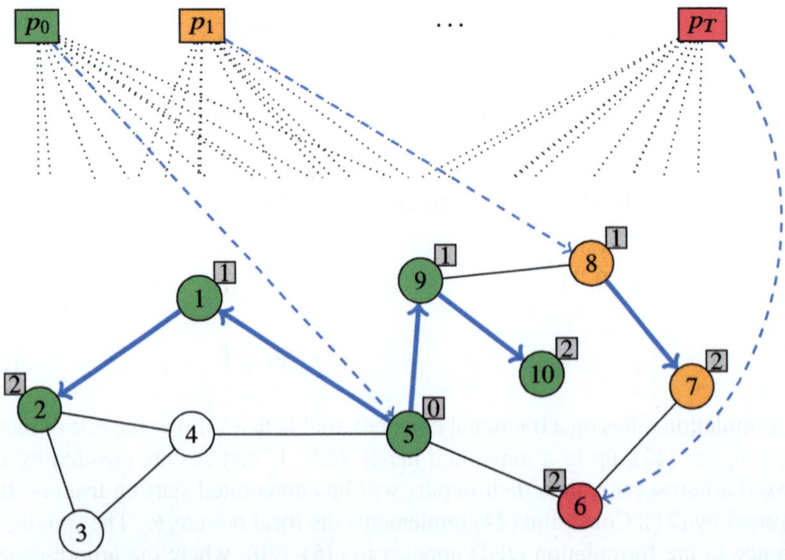

Fig. 1 An illustration of the MTZ formulation for $V = \{1, \ldots, 10\}$, $\theta_v := \lfloor \deg(v)/2 \rfloor$, and $T = 2$. Depth-trackers are displayed as attached rectangular labels. Association to the same arborescence is indicated by the same vertex color, activating edges by arrows.

4 A Heuristic Approach

As we will see in the computational results, even the most adequate MILP approach reached its time limit on θ-GBP instances with thousands of vertices due to the high number of activation and depth-encoding variables. Thus, we propose an alternative heuristic approach relying on a simplified assumption. It will be able to produce solutions of solutions of satisfactory quality for large-scale instances requiring shorter runtimes. A typical greedy approach in the setting of the θ-GBP would be to iteratively find arborescences of decreasing heights and let them cover the input graph. After the placement of an arborescence, all used vertices are eliminated from the working copy of the input graph before the next arborescence is derived with the bound on its height decremented by one. The procedure terminates when the height bound reaches zero. Note that it may happen that all vertices are already covered earlier, in which case we assume here that all remaining arborescences are empty. The core task in this heuristic framework concerns the solution of the intermediate one-arborescence problem MCADH(G, θ, h), as introduced in Sect. 2.

The first approach relies on a straightforward adaptation of the ILP (1)–(12) addressing the search of a single arborescence; here, we can eliminate many activation variables per edge. We refer to this approach as *Greedy One Arborescence at a Time via Integer Linear Programming* (GOAT-ILP).

Recalling that many heuristic approaches in the literature, e.g., [14, 15] rely on the eigencentrality, we sketch a respective approach in our setting. The idea is to concurrently achieve the placement of a single arborescence via the iterative extension of an initial arborescence A (initially consisting of solely a root vertex) by finding the most promising current leaf w of A and strategically appending to it as many further vertices as the threshold of w allows. This growing process is guided by a scoring function quantifying how useful it is to append to a leaf w of A a neighbor $z \in N(w)$ (according to the adjacency of the original graph G) which is not yet present in the arborescence. We use the scoring function

$$\mathsf{LN}(A, w; z) := d(w) \cdot \frac{\theta'(z)}{\max_{y \in V} \theta'(y)} + (T - 1 - d(w)) \cdot \mathrm{ec}_{G-(A \cup \{w,z\})}(z),$$

where $d(w)$ denotes the depth of w inside A. The value $\theta'(z)$ refers to the remaining effective threshold of vertex z taking into account that some neighbors of z may already be incorporated into the arborescence; thus this threshold for the count of neighbors that might be ignited via diffusion can already be smaller. Lastly, $\mathrm{ec}_{G-(A \cup \{w,z\})}(z)$ denotes the eigencentrality of z after excluding the edge $\{w, z\}$ and the vertices of the so-far built arborescence A from G.

The (renormalized) value of $\theta'(z)$ and the eigencentrality of z—both attaining values in the interval $[0, 1]$—are linearly combined by weights assuring that leaves whose current height is far from the maximum height T are increasingly rewarded by a high eigencentrality value, and gradually, as the depths of the leaves increase, their (renormalized) θ'-value receives more importance. In the situation just before the depth reaches the bound on the arborescence's height, by design, the eigencentrality does not even contribute to LN.

Eventually we want to find a leaf w of A for which

$$\mathsf{LeafScore}(A, w) := \max_{\substack{P \subseteq N(w) \setminus A \\ |P| \leq \theta'(w)}} \sum_{z \in P} \mathsf{LN}(A, w; z) \tag{29}$$

is maximal among the leaves of A. When this maximizer w is found, the arborescence A is extended towards all vertices in the set P that was itself responsible for maximizing (29).

We start the whole construction by initializing A with a vertex having highest eigencentrality, apply aforementioned procedure as long as an extension is possible, and return this maximal arborescence as result for the MCADH instance. We refer later to this score-based approach carried out iteratively for $t = T, T - 1, \ldots, 1, 0$, simply as *Greedy One Arborescence at a Time* (GOAT).

5 Computational Results

The MILP and GOAT approaches proposed in Sect. 3 have been implemented in Julia 1.11.1. The MILP solver Gurobi[2] in version 10.0.3 is used to solve the MILPs. A cluster with an Intel(R) Xeon(R) E5-2640 v4 CPU with 2.40GHz and 160GB RAM running Ubuntu 18.04.6 LTS was used to run all experiments on a single thread.

We rely on the benchmark instances from [17, Table 3]. Here, we show results only for every fourth instance from this benchmark set due to space limitations. The results for all instances can be found online.[3] As thresholds we use $\theta_v := \lfloor v/2 \rfloor$, $v \in V$, comparable to the so-called *majority thresholds* for the Target Set Selection Problem [2]. The following approaches are considered in the comparisons: OHD; MTZ; MTZ-s, i.e., MTZ strengthened by the distance-based constraints (28); GOAT-ILP with a time limit of 1000 seconds per placement; and GOAT which will be let run until termination without a set timelimit. A time limit of 1000 seconds is given to Gurobi to solve OHD, MTZ, MTZ-s. No time limit is set for the original GOAT approach. For each approach after termination an independently written feasibility-checker is run.

Table 1 is structured as follows: each column corresponds to a selected instance, labeled with its respective name. The first three rows provide key characteristics of each instance, including the number of vertices, the number of edges, and the standard deviation of vertex degrees. The subsequent rows present the results for OHD, MTZ, MTZ-s, GOAT-ILP, and GOAT, evaluated for $T \in \{2, 3\}$. For the OHD and the GOAT approach, the runtimes in seconds are additionally documented; see rows OHD-time[s] and GOAT-time[s], respectively. Gaps in percent shown in Table 1 indicate the difference between the dual bound and the primal bound, further renormalized by the primal bound. The gaps for MTZ and MTZ-s, are omitted here, as they often arise just from the trivial dual bound, i.e., the number of vertices. The following observations can be made from the reported results in Table 1.

For $T = 2$ the OHD approach is able to solve all instances to optimality where the used time is shorter than the time limit of 1000 seconds—often under a fraction of a second. We notice that none of the other exact approaches, i.e., MTZ and MTZ-s, is able to return results of comparable quality; they just can solve the two sparsest instances sphere3 and lattice2D (almost) optimally. When examining the slightly denser instances, already for polbooks with only 105 vertices both MTZ approaches do not find a primal bound of at least 90% of the optimal one computed by OHD despite a 200 times longer runtime. This trend is even more evident on larger, dense instances where the incumbent solution of the MTZ approach can be smaller by (almost) an order of magnitude (see ia-fb-messages and TVshow). On the instances with up to 500 vertices, the additional distance-based constraints in MTZ-s seem indeed to help finding better solutions than those obtained via the MTZ formulation. On the other hand, for larger graphs the numerous strengthening

[2] https://www.gurobi.com.

[3] https://www.ac.tuwien.ac.at/research/problem-instances/#Graph_Burning_Problem.

Table 1 Comparison of approaches for $T \in \{2, 3\}$. Rows OHD report optimal values except for $T = 3$ with encountered gaps of 426% and 12% for `c-fat500-1` and `TVshow`, respectively.

	karate-club	polbooks	ia-enron-only	sphere3	c-fat500-1	web-polblogs	DD199	lattice3D	lattice2D	ia-fb-messages	TVshow
$\|V\|$	34	105	143	258	500	643	841	1000	1089	1266	3892
$\|E\|$	78	441	623	768	4459	2280	1902	2700	2112	6451	17262
sd(deg)	3.88	5.47	6.08	0.30	1.31	11.46	1.47	0.69	0.34	13.24	12.6
$T = 2$											
OHD	34	76	122	18	47	369	30	18	11	778	753
MTZ	34	68	90	18	41	293	19	12	11	163	84
MTZ-s	34	69	118	18	46	190	9	12	10	5	0
GOAT-ILP	31	76	122	18	47	369	30	18	11	761	753
GOAT	25	62	109	18	47	283	26	18	11	697	204
OHD time[s]	0.06	4.41	1.20	0.66	127.2	20.41	1.17	2.37	1.66	41.17	231.0
GOAT time[s]	0.01	0.01	0.02	0.01	0.03	0.24	0.02	0.01	0.02	21.73	7.52
$T = 3$											
OHD	34	105	143	52	95	612	68	58	26	1237	1861
MTZ	34	105	143	46	82	513	36	33	26	629	100
MTZ-s	34	105	143	48	86	587	23	43	16	1108	0
GOAT-ILP	34	105	139	52	96	583	67	58	26	1235	1807
GOAT	21	99	138	47	89	402	55	56	26	1062	626
OHD time[s]	0.01	0.15	0.29	8.19	1000	21.68	20.13	0.24	11.98	67.13	1000
GOAT time[s]	0.01	0.03	0.06	0.02	0.08	0.67	0.08	0.07	0.07	42.88	30.73

inequalities negatively impact, apparently, on the solution quality yielding solutions smaller by even two orders of magnitude for the two largest graphs. The results for the GOAT-ILP indicate that the successive arborescence placement seems to be a promising heuristic. Despite the considerably simplifying nature of the approach, primal bounds on par with (or just slightly inferior to) the optimal primal bounds from OHD can be obtained. On smaller graphs the GOAT heuristic is slightly inferior to the exact approaches. However, on graphs with more than 800 vertices it outperforms the MTZ approaches. While the quality determined by the OHD approach cannot be obtained by GOAT, the computation time of the latter is in general shorter.

When $T = 3$ we notice that runtimes for OHD are mostly longer than for $T = 2$, which is explainable by the increased number of variables in the respective formulation. On the three smallest instances the runtime is, however, shorter than for $T = 2$, which is most likely due to the fact that a solution burning the entire graph is now quickly encounterable, making the computation of tight dual bounds unnecessary. Apparently, for the instance `c-fat500-1` with $T = 3$, the calculation of tight dual bounds is particularly difficult in the modality of OHD; we notice that GOAT-ILP is in this case the preferable approach returning a solution better by one unit. On the remaining instances GOAT-ILP has again a quality on par or slightly inferior to OHD.

For $T \in \{2, 3\}$ the GOAT approach does not appear to fully exploit the quality that can be obtained according to GOAT-ILP in the greedy setting, but it is superior to the MTZ approaches on the largest instance; the MTZ approaches seem to be unsuitable for processing such large graph sizes in a meaningful way. In contrast to the exact ILP-driven routine in GOAT-ILP, the score function driven placement in GOAT risks leading the arborescence towards dead ends or locally unpromising vertices. However, in particular if the need is to quickly generate a solution, this seems a viable approach given the faster—and apparently better estimable—computation times.

It is an important insight that, when dealing with a large number of binary variables in OHD, the solver performs better than in the setting of MTZ where many of these variables are replaced by fewer fractional variables.

6 Conclusion

We introduced a generalized version of the well-studied graph burning problem that more realistically enables control of the diffusion process via local threshold parameters. During diffusion it limits the number of neighboring vertices to be possibly affected by a burned vertex in the subsequent iteration. This problem is linked to the task of finding a shortest sequence of vertex-disjoint arborescences with upper-bounded heights and local upper bounds on their out-degrees.

Following this observation, we proposed two MILP models, one of which was clearly ahead on the instances tested. Empirically shown, the addition of a greedy assumption has proven its worth with only minimal loss of quality counterbalanced by a clear simplification of the problem. The availability of a suitable local search would possibly add the finishing touches to near-optimal solutions, in particular for GOAT-ILP.

Our greedy approaches still neglect the aspect of "communication" between the different arborescences. In future work, one may address their simultaneous construction. We plan to let a score function also consider the connectivity, in particular the number and topology of the arising connected components, to enhance the approach.

Furthermore, several open problems arise in the context of the introduced θ-burning process: From the perspective of fire spread, the θ-burning number expresses a worst-case that we are interested in precalculating. However, it might be also relevant to consider the setting where a number of at most θ_v unburned neighbors picked uniformly at random are chosen to receive the status burned—at the same time, we might want to assume a random choice of the seeds. This yields a different measure of vulnerability and coming up with estimates for the expected penetration over all processes obeying these assumptions seems an interesting challenge.

The following bilevel optimization problem may be a relevant extension: Given a budget $B \in \mathbb{N}$, find a vector of "threshold-reducers" $(r_v)_{v\in V}$ with $0 \leq r_v \leq \theta_v$ and $\sum_{v\in V} r_v \leq B$ such that for $\theta' := (\theta_v - r_v)_{v\in V}$ the respective value of $b_{\theta'}$ is minimum, i.e., find a budget-maintaining way to contain the worst-case of fire diffusion as much as possible.

Acknowledgements We thank an anonymous referee and Johannes Varga for valuable inputs. This work is partially supported by the VGSCO (FWF [10.55776/W1260-N35]), the Ministry of Civil Affairs of Bosnia and Herzegovina [1259074], and OeAD [BA05/2023]; M. Djukanović further by the Ministry of Scientific and technological development and higher education, Serb Republic ["Utilizing deep learning to enhance the efficiency of optimization algorithms"].

References

1. Bonato, A., Janssen, J., Roshanbin, E.: Burning a graph as a model of social contagion. In: Algorithms and Models for the Web Graph: 11th International Workshop. LNCS, vol. 8882, pp. 13–22. Springer (2014)
2. Kempe, D., Kleinberg, J., Tardos, É.: Maximizing the spread of influence through a social network. In: Proceedings of the 9th ACM SIGKDD International Conference on Knowledge Discovery and Data Mining. pp. 137–146. Association for Computing Machinery (2003)
3. Günneç, D., Raghavan, S., Zhang, R.: Least-cost influence maximization on social networks. INFORMS J. Comput. **32**(2), 289–302 (2020)
4. Bessy, S., Bonato, A., Janssen, J., Rautenbach, D., Roshanbin, E.: Burning a graph is hard. Discret. Appl. Math. **232**, 73–87 (2017)
5. Hiller, M., Koster, A.M., Triesch, E.: On the burning number of p-caterpillars. In: Graphs and Combinatorial Optimization: from Theory to Applications, AIRO, vol. 5, pp. 145–156. Springer (2021)
6. Liu, H., Hu, X., Hu, X.: Burning number of caterpillars. Discret. Appl. Math. **284**, 332–340 (2020)
7. Bonato, A.: A survey of graph burning. Contrib. Discret. Math. **16**(1), 185–197 (2021)

8. Norin, S., Turcotte, J.: The burning number conjecture holds asymptotically. J. Comb. Theory Ser. B **168**, 208–235 (2024)
9. Murakami, Y.: The burning number conjecture is true for trees without degree-2 vertices. Gr. Comb. **40**(4), 82 (2024)
10. Bonato, A., Kamali, S.: Approximation algorithms for graph burning. In: Theory and Applications of Models of Computation: 15th Annual Conference. LNCS, vol. 11436, pp. 74–92. Springer (2019)
11. Lieskovský, M., Sgall, J., Feldmann, A.E.: Approximation algorithms and lower bounds for graph burning. In: Approximation, Randomization, and Combinatorial Optimization. Algorithms and Techniques. LIPIcs, vol. 275, pp. 9:1–9:17. Schloss Dagstuhl—Leibniz-Zentrum für Informatik (2023)
12. García-Díaz, J., Pérez-Sansalvador, J.C., Rodríguez-Henríquez, L.M.X., Cornejo-Acosta, J.A.: Burning graphs through farthest-first traversal. IEEE Access **10**, 30395–30404 (2022)
13. Farokh, Z.R., Tahmasbi, M., Tehrani, Z.H.R.A., Buali, Y.: New Heuristics for Burning Graphs. arXiv preprint arXiv:2003.09314 (2020)
14. Gautam, R.K., Kare, A.S.: Faster heuristics for graph burning. Appl. Intell. **52**(2), 1351–1361 (2022)
15. Šimon, M., Huraj, L., Dirgová Luptáková, I., Pospíchal, J.: Heuristics for spreading alarm throughout a network. Appl. Sci. **9**(16), 3269 (2019)
16. García-Díaz, J., Cornejo-Acosta, J.A., Trejo-Sánchez, J.A.: A greedy heuristic for graph burning. Computing **107**(3), 91 (2025)
17. García-Díaz, J., Rodríguez-Henríquez, L.M.X., Pérez-Sansalvador, J.C., Pomares-Hernández, S.E.: Graph burning: mathematical formulations and optimal solutions. Mathematics **10**(15), 2777 (2022)
18. de Carvalho Pereira, F., de Rezende, P.J., Yunes, T., Morato, L.F.B.: A row generation algorithm for finding optimal burning sequences of large graphs. In: 32nd Annual European Symposium on Algorithms. LIPIcs, vol. 308, pp. 94:1–94:17. Schloss Dagstuhl—Leibniz-Zentrum für Informatik (2024)
19. Wu, F., Huberman, B.A.: Novelty and collective attention. Proc. Natl. Acad. Sci. **104**(45), 17599–17601 (2007)
20. Finbow, S., MacGillivray, G.: The firefighter problem: a survey of results, directions and questions. Australas. J. Comb. **43**, 57–78 (2009)
21. Miller, C.E., Tucker, A.W., Zemlin, R.A.: Integer programming formulation of traveling salesman problems. J. ACM **7**(4), 326–329 (1960)

A Robust Approach for the Prepositioning of Resources for Wildfire Suppression

Francisco Marques, Agostinho Agra, Helena Alvelos, Ana Raquel Xambre, and Filipe Alvelos

Abstract We consider the problem of prepositioning a set of firefighting resources for wildfire suppression. As wildfires are highly affected by uncertainty we devise a two-stage model where the prepositioning of resources are first-stage decisions and the movement of these resources after the fire ignitions are known are the second-stage decisions. We use the minimum travel time principle and mixed integer programming to model the fire spread in the landscape and decisions related to the prepositioning of resources and their movement to attack positions. To model uncertainty we use robust optimization based on a discrete set of scenarios which represent ignition locations, wind speed and directions. As the size of the model grows considerably with the number of scenarios, a row-and-column decomposition algorithm is proposed. Computational experiments based on an actual landscape are reported showing the efficiency of the decomposition algorithm.

Keywords Wildfire suppression · Prepositioning of resources · Robust optimization · Decomposition algorithm

F. Marques
University of Aveiro, Aveiro, Portugal
e-mail: franciscocmarques@ua.pt

A. Agra (✉)
Departamento de Ciências Matemáticas and Centro de Estudos Matemáticos (CEMS.UL), Faculdade de Ciências, Universidade de Lisboa, Lisboa, Portugal
e-mail: amagra@fc.ul.pt

H. Alvelos · A. R. Xambre
Department of Economics, Management, Industrial Engineering and Tourism/Center for Research and Development in Mathematics and Applications, University of Aveiro, Aveiro, Portugal
e-mail: helena.alvelos@ua.pt

A. R. Xambre
e-mail: raquelx@ua.pt

F. Alvelos
Department of Production and Systems/ALGORITMI Research Center/LASI, School of Engineering, University of Minho, Braga, Portugal
e-mail: falvelos@dps.uminho.pt

F. Alvelos et al. (eds.), *Advances in Optimization and Wildfire*, Lecture Notes in Operations Research, https://doi.org/10.1007/978-3-032-03108-2_10

1 Introduction

Wildfires are natural disasters that can have a significant impact on ecosystems which can lead to the loss of lives and material damage. Therefore, their management is a highly relevant issue. Following [1], wildfire management can be divided in three main areas: prevention, detection and management of fire suppression resources. In this paper, we address an operational management of fire suppression resources problem. Although the first operational research approaches to this type of problems are from the early 1960s [2], only recently a categorized literature review [3] and a survey [4] have been published, supporting the current relevance of the topic.

Operational management of fire suppression resources can be typified in deployment, dispatch and positioning. Deployment decisions are taken before an actual fire is detected and typically amount to defining which resources should be (re)allocated to which bases. Dispatch decisions are taken after an ignition is detected and amount to deciding which resources should be sent to the fire. Lastly, positioning decisions detail where the dispatched resources should attack the fire.

In this paper, we consider a problem that combines the prepositioning and positioning of resources. The decisions of repositioning the resources, which can be seen as deployment decisions but with allocation alternatives other than the bases—for example, roads—are taken before fire is detected. After a fire is detected, the attack positions of the resources are defined. Given the uncertainty involved in the fire occurrence (and also on the wind), we derive an optimization under uncertainty approach based on mixed integer programming (MIP).

Different MIP models have been proposed for the positioning problem. In [5], the minimum travel time principle (MTT, [6]) was first used to model the fire paths and arrival times in a landscape. According to this principle, fire follows the quickest path between any two locations of a landscape. This principle allows to model fire spread as shortest paths in a network and has since been used in software that simulates fire spread (e.g. [7]) and in MIP models for different variants of the positioning problem, for example in [3, 8–10], and [11]. These are deterministic approaches that, at the expense of increasing the complexity and difficulty of the approach, can be extended to address uncertainty.

The two main stream optimization approaches to deal with uncertainty are stochastic optimization (SO) [12] and robust optimization (RO) [13, 14]. While stochastic optimization assumes that the probability distribution of the uncertain parameters are known, the robust optimization only assumes that the parameters belong to an uncertainty set. In addition, while SO optimizes over the expected value, robust optimization evaluates the decisions under the worst-case scenario. On wildfire, examples of a SO approach is [15] (extending [9]), where a multi-stage stochastic programming is proposed to address wind uncertainty. An example of RO is [16], where a positioning problem with uncertainty in the number of resources required to suppress fire in a given position is addressed (extending [3]).

In this paper, we develop a MIP model, based on [10], to address uncertainty on the location of the ignitions and wind. We consider that the spatial distribution of fuels and slopes of the landscape under study are known. Wildfire uncertainty is modelled through a set of scenarios, each one defined by one or more ignitions and a wind direction and intensity. The problem is to decide where to prepositioning the resources such that the burned area is minimized in the worst-case scenario. This approach can be seen as having static variables (related to the prepositioning decisions) and adjustable variables (related to the decisions of each scenario). Both the number of constraints and the number of adjustable variables depend on the number of scenarios. Therefore, to solve the problem efficiently we propose a row-and-column decomposition algorithm. A decomposition approach was first introduced in an inventory model in [17]. Later, it was independently, introduced in [18] for a robust vehicle routing problem and as a general approach in [19].

This paper is organized as follows: we provide a description of the problem in Section 2. In Section 3, we introduce the RO model. The row-and-column decomposition algorithm is presented in Section 4. The computational experiments conducted to provide insights on the solutions and to evaluate the proposed model and decomposition algorithm are reported in Section 5. Finally, in Section 6, we provide the conclusions and future research directions.

2 Problem Definition and Notation

We consider the problem of prepositioning a set of available resources, possibly of different types, in a landscape before a wildfire ignition occurs. The objective is to minimize the worst-case burned area, within a given time period, for a set of scenarios. Each scenario is characterised by one or more ignitions and a wind speed and direction. After the prepositioning decision is taken and the uncertainty is disclosed (i.e. the actual scenario is known), the resources can be moved from their prepositioning locations to the ones where their impact on the burned area is greater.

In the context of optimization under uncertainty, the problem considered is a two-stage problem. In the first-stage, the prepositioning of the resources is decided, i.e. static decisions are taken. Then the uncertainty (ignition(s) locations and wind) is revealed. In the second-stage, the (adjustable) decisions on the movement from prepositioning positions to the positions where the resources attack fire are taken.

In order to model this problem, we represent the landscape as a graph, denoted by the fire graph, where the nodes are associated with the hexagons of a grid obtained by discretizing the landscape, and the arcs are associated with the potential transmission of fire between two adjacent nodes. Each arc has a fire transmission time that corresponds to the time it takes for the fire to reach the destination node of the arc once the fire reaches the origin node. In our computational experiments, the fire transmission time of the arcs was derived from the fire behaviour model of Rothermel [20] that makes use of fuels, slopes and winds to estimate the fire rate of spread in homogeneous conditions. It is worth noting that, in the optimization model, the fire transmission time is a parameter and therefore can be obtained with other methods.

Based on the fire transmission times, it is possible to model the fire spread by using the minimum travel time principle that states that the time it takes to the fire to reach a node is the length (duration) of the shortest (quickest) path from an ignition node.

We now introduce the notation used. We consider a set of available resource types defined by R. Each resource type, r, $r \in R$, has associated a graph, represented by $G^r(N^r, A^r)$, where the nodes correspond to potential attack positions and the arcs to potential movements. Each resource type r, $r \in R$, also has a number of available units, b^r.

The fire graph, already described, is represented by $G^f(N^f, A^f)$. The fire and resource nodes are related through a set of pairs (l, i), denoted by A^{rf}, where l is a node of the resource network and i is a node of the fire graph. In this work, we set A^{rf} is based on pairs of nodes with a distance below a given threshold.

If a unit of resource r is positioned at a node l of its resource network, then all the nodes i of the fire network for which $(l, i) \in A^{rf}$ are attacked. A fire node being attacked means that the fire transmission time of all the outgoing arcs are increased by a constant Δ^r. This constant models the suppression productivity. Although in practice it may depend on several factors, e.g. the flame length and fire intensity, in this work we assume it is constant for each resource type.

The instant where the burned area is evaluated is denoted by T. The constant d^r_{kl} is the quickest path duration of one unit of resource r between nodes kl. There is always a time gap between the ignition and the time the fire is detected. We define this gap as the alert time T_a. This duration is then added to the quickest path duration of the resources, d^r_{kl}, resulting in the actual time that each unit takes to reach a location. We represent the set of nodes that are burning at the alert time, e.g. the fire perimeter, as I^f. The fire transmission time between adjacent nodes ij in scenario ω is indicated by c^{ω}_{ij}.

Figures 1 and 2 depict an example of the positioning of resources and the fire spread, using 5 and 10 units, respectively. When the number of available units is higher, it becomes clear, in Figure 2 the effectiveness of the presence of the resources in delaying the fire spread and reducing the number of burned nodes.

3 Robust Model

The MIP model that address the robust prepositioning problem has the following static (first-stage) decision variables:

$$z^r_k = \begin{cases} 1, & \text{if a unit of a resource of type } r \text{ is prepositioned at node } k \\ 0, & \text{otherwise} \end{cases}$$

$$k \in N^r, r \in R.$$

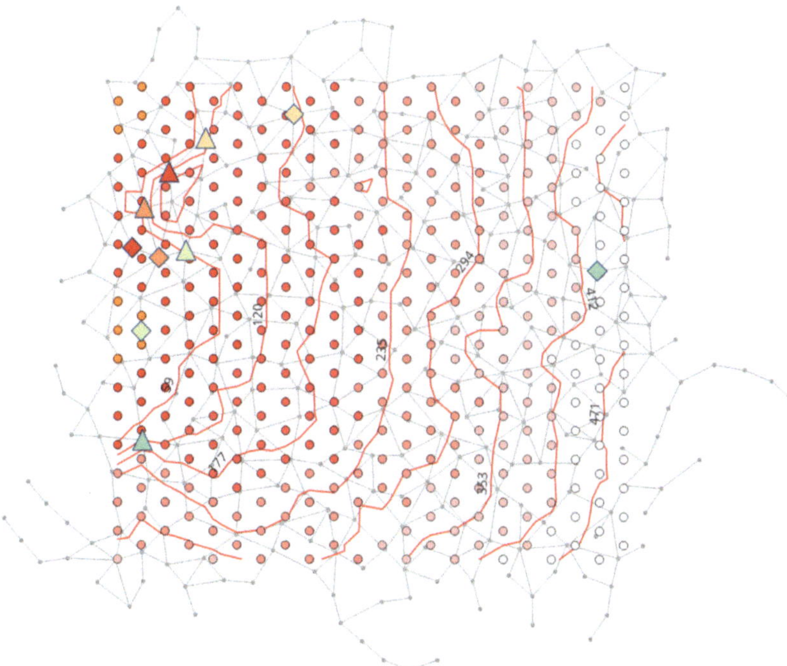

Fig. 1 Example of an optimal resource prepositioning and positioning in the worst-case scenario obtained by the robust model. The orange nodes are the ignition nodes, and the fire network nodes are represented by a fading red based on their shortest distance to the ignition nodes. The red lines are isochronous relative to the fire arrival time, in minutes. The coloured diamonds are the prepositioning nodes, and the coloured triangles are the positioning of the resources in the second stage. The grey nodes and arcs are the roads of the landscape. Worst-case scenario of the two scenarios that were used in instance 21 (see computational experiments section below), with ignitions in W, and (W, NW). The wind attributes were *upslope* and speed of 30*km*/*h*. Five resource units available

To model uncertainty we consider a set of scenarios, denoted by Ω. A single scenario represents a set of ignition locations, wind speed and direction, and will be denoted by ω. For each scenario $\omega \in \Omega$ we define the following adjustable (second-stage) decision variables:

t_i^ω fire arrival time at node i in scenario ω, $i \in N^f$, $\omega \in \Omega$;

$$x_{kl}^{\omega r} = \begin{cases} 1, & \begin{array}{l}\text{if a unit of a resource of type } r \text{ is prepositioned at node } k\\ \text{is moved to node } l \text{ in scenario } \omega\end{array} \\ 0, & \text{otherwise} \end{cases}, kl \in A^r, r \in R, \omega \in \Omega;$$

$$y_i^\omega = \begin{cases} 1, & \text{if node } i \text{ burns before instant } T \text{ in scenario } \omega \\ 0, & \text{otherwise} \end{cases}, i \in N^f, \omega \in \Omega.$$

The robust problem can be modelled as follows:

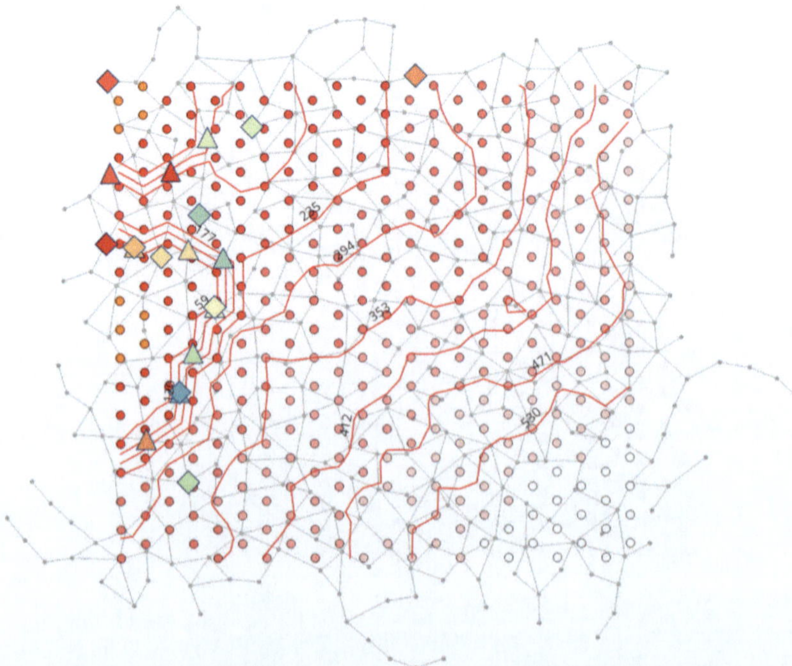

Fig. 2 Illustration similar to Figure 1 but for the Worst-case scenario of the two scenarios that were used in instance 22 (see computational experiments section below), with ignitions in W, and (W, NW), wind attributes given by speed of 30*km/h* and direction *upslope*, and ten resource units

$$\min_{(z,t(.),x(.),y(.))\in X} \max_{\omega\in\Omega}\left(\sum_{i\in N^f} y_i^{\omega}\right)$$

where X is the feasible set defined by a set of constraints.

In order to devise a MIP (linear) model we use the well-known epigraph formulation to remove the inner maximization problem. Let η be the number of burned nodes in the worst-case scenario:

$$\eta = \max_{\omega\in\Omega} \sum_{i\in N^f} y_i^{\omega}. \tag{1}$$

The MIP model is the following:

$$\text{Min } \eta \tag{2}$$

$$\text{s.t. } \eta \geq \sum_{i\in N^f} y_i^{\omega},\ \forall \omega \in \Omega \tag{3}$$

$$\sum_{k \in N^r} z_k^r \leq b^r, r \in R \tag{4}$$

$$z_k^r \in \{0, 1\}, k \in N^r, r \in R \tag{5}$$

$$t_i^\omega = 0, i \in I^f, \omega \in \Omega \tag{6}$$

$$t_j^\omega \leq t_i^\omega + c_{ij}^\omega + \sum_{r \in R} (\Delta^r \sum_{\substack{kl \in A^r, \\ li \in A^{rf}}} x_{kl}^{\omega r}), ij \in A^f, \omega \in \Omega \tag{7}$$

$$(d_{kl}^r + T_a) x_{kl}^{\omega r} \leq t_i^\omega, kl \in A^r, li \in A^{rf}, i \in N^f, r \in R, \omega \in \Omega \tag{8}$$

$$\sum_{l \in N^r} x_{kl}^{\omega r} = z_k^r, \forall k \in N^r, r \in R, \omega \in \Omega \tag{9}$$

$$y_i^\omega \geq 1 - t_i^\omega / T, i \in N^f, \omega \in \Omega \tag{10}$$

$$t_i^\omega \geq 0, i \in N^f, \omega \in \Omega \tag{11}$$

$$x_{kl}^{\omega r} \in \{0, 1\}, kl \in A^r, r \in R, \omega \in \Omega \tag{12}$$

$$y_i^\omega \in \{0, 1\}, i \in N^f, \omega \in \Omega \tag{13}$$

The objective function (2) minimizes the worst-case in the set of scenarios. Notice that Constraints (3) impose the lower bound of η as the sum of burned nodes, for every scenario, thus for the worst-case scenario. As the objective function is a minimization problem it forces that in the optimal solution η would not be greater than that maximum. Constraints (4) limit the number of resource units for each resource type. Constraints (6) define the nodes of the fire initial perimeter, that is, the burning nodes when the fire is detected. Constraints (7) define the fire spread: the fire arrival time at each node is the minimum of the potential fire arrival times from each of the incoming arcs. A resource in the origin of an arc delays the fire transmission. Constraints (8) state that a resource can only be moved to a node if the resource manages to reach the node before the fire. Constraints (9) state that a resource can only be moved from a position if it was prepositioned in that location. Note that $x_{ll}^{\omega r}$ corresponds to keeping the resource in the same position. Due to the domain of z_k^r, there can only be one resource of type r per node. Constraints (10) relate the fire arrival time with the node being burned or not at instant T. The remaining constraints (5), (11), (12), (13) define the domain of the variables.

4 Row and Column Decomposition Algorithm

As solving the robust MIP directly can be computationally expensive because the size of the model increases with the number of scenarios (and each scenario can translate itself in a large number of decision variables and constraints), we apply a row and column decomposition algorithm. This algorithm solves, iteratively, a master problem and a subproblem. The master problem, *RobustMaster*(S), is the robust model for a subset of scenarios S (initially the first scenario). By solving the master problem we obtain the first-stage solution $\bar{z}$ (optimal prepositioning nodes for the set of scenarios S) and the corresponding worst-case burned area, η. The subproblem, *RobustSub*($\bar{z}$), takes as input the fixed prepositioning $\bar{z}$ obtained by the master problem, and computes the sum of burned nodes in each scenario ω not in S, θ_ω. This subproblem solves $\mid \Omega \setminus S \mid$ deterministic problems, one for each scenario. If a scenario ω leading to a burned area $\theta_\omega > \eta$ is found, then the corresponding variables and constraints are added to the master problem and the process is repeated. Otherwise, if no such scenario is found, then the algorithm stops.

The full decomposition algorithm is given in Algorithm 1.

Algorithm 1: Row and column decomposition algorithm

```
Stop ← 0 ;
S ← {ω1};
repeat
    Solve RobustMaster(S);
    Let z̄ be the optimal prepositioning for the set of scenarios S and η be the
      number of burned nodes in the worst-case;
    θ ← 0;
    ω' ← 0;
    forall ω ∈ Ω\S do
        Solve deterministic problem for scenario ω and prepositioning z̄;
        Let θ_ω be the number of burned nodes in scenario ω;
        if θ_ω > θ then
            θ ← θ_ω;
            ω' ← ω
        end if
    end forall
    if θ > η then
        S ← S ∪ {ω'};
    else
        Stop ← 1
    end if
until Stop;
```

Algorithm 1 Row and column decomposition algorithm

5 Computational Results

The computational tests were performed on a 8th Gen i7 Intel CPU with 1.8GHz. The optimization models were implemented in Python using the Gurobi solver version 10.0.2 and pyO3F (a Python framework being developed under the scope of the project "An optimization framework for reducing forest fire") [21, 22] that allows to define from an actual landscape, the fire and resources networks for a given resolution and fire spread scenarios depending on the fuels, slopes and winds. The computational experiments were conducted for a landscape surrounding the village of Baião, a village on the north of Portugal, with around 20 km by 18 km (an extent of around 360 km^2). The resolution (distance between two adjacent nodes of the hexagonal grid representing the landscape) used was 1000 metres. In all of these tests we used $T_a = 30$ minutes and $T = 120$ minutes. Only one type of resource was used with different speeds based on the type of road that they are on: on the primary and secondary roads of $60 km/h$ and on paths not suitable for vehicles, $5 km/h$. Each resource unit has a suppression ray of 1000 metres, i.e. every fire node within this distance of a resource unit is attacked.

We tested the two methods, direct and decomposition-based, in 22 instances. Each of these instances is made up of 3 attributes: the ignitions in each scenario, the landscape wind properties, and the number of available resource units. The number of scenarios in each instance is obtained from the combinations of the ignitions with the wind properties. The fire rate of spread in each arc is obtained based on the slope, fuel and wind. Slope and fuel are common to all instances and obtained directly from public data and processed by pyO3F (details can be found in [23]).

The possible ignition nodes are defined by the landscape nodes closest to each cardinal point (N, NW, W, SW, S, SE, E, NE) and to the centre (C). The wind has two modes: *upslope* and *fixed*. The *Upslope* mode uses the wind direction that increases the fire spread at each node of the grid the most (i.e. in the direction of the slope), and the *fixed* mode uses a fixed direction throughout all of the fire spread arcs (the wind is projected according to [24]). Using *fixed* wind direction, 0° means that the wind is moving from south to north and the direction is clockwise.

Table 1 describes the attributes of each instance. The columns used in this table are: Instance—the index of the instance; Available Units—the number of available resource units; Ignitions—a list containing all ignition locations in each scenario (ignitions inside brackets indicate scenarios with more than one ignition); Wind Properties—the attributes of the wind in the scenarios.

Table 2 displays the results obtained using the robust model (using the default solver options) and the row-and-column decomposition algorithm, and Table 3 presents the performance metrics of these two approaches. The new columns used in the Tables are:

- Units—number of available resource units for use, and how many were used in the fire suppression;
- Burned Nodes—maximum and average number of burned nodes in the set of scenarios;

Table 1 Description of the instances that were used on computational experiments

Instance	Available Units	Ignitions	Wind properties Mode, Speed (*km/h*), Direction (°)
1	5	N, NW	Upslope, 30, -
2	5	N, NW, W	Upslope, 30, -
3	5	N, NW, W, SW	Upslope, 30, -
4	5	N, NW, W, SW, S	Upslope, 30, -
5	5	N, NW, W, SW, S, SE	Upslope, 30, -
6	5	N, NW, W, SW, S, SE, E	Upslope, 30, -
7	5	N, NW, W, SW, S, SE, E, NE	Upslope, 30, -
8	5	N, NW, W, SW, S, SE, E, NE, C	Upslope, 30, -
9	5	NE	Upslope, 30, -
10	5	NE	Fixed, 30, 30
11	5	NE	Fixed, 60, 30
12	5	NE	Fixed, 60, 60
13	5	NE, C	Upslope, 30, -
14	5	NE, C	Fixed, 30, 30
15	5	NE, C	Fixed, 60, 30
16	5	NE, C	Fixed, 60, 60
17	2	NE, E, (NE, E), (NE, E, SE)	Fixed, 30, 10
18	2	NE, E, (NE, E), (NE, E, SE), (NE, C, E)	Fixed, 30, 10
19	2	NE, E, (NE, E), (NE, E, SE)	(Fixed, 30, 10), (Fixed, 20, 210)
20	2	NE, E, (NE, E), (NE, E, SE), (NE, C, E)	(Fixed, 30, 10), (Fixed, 20, 210)
21	5	W, (W, NW)	Upslope, 30, -
22	10	W, (W, NW)	Upslope, 30, -

- Runtime—duration of the optimization of the implemented solution procedure in seconds. For the decomposition approach, the total durations of the master and subproblems are also recorded separately;
- Iterations—number of times that the master problem was solved to reach the optimal solution of the set of scenarios.

From Table 2 it is clear that both models are successful in delaying the fire spread. In Table 2, there are some instances where the number of used resources is 0 due to the slow fire spread. Since no nodes, other than the ignition node, burn until time T, when the number of burned nodes is evaluated, the use of resources does not have any impact.

In instances 17–20, the number of available resources was decreased from 5 to 2 to study the effects of having a small number of fire fighting units. We can observe that the objective function values coincide in both approaches, as was expected. However,

Table 2 Computational results of the robust models

	Robust model				Decomposition algorithm		
Instance	Units		Burned Nodes		Units	Burned Nodes	
	Available	Used	Max	Mean	Used	Max	Mean
1	5	5	12.0	12.0	5	12.0	12.0
2	5	5	36.0	35.67	5	36.0	33.5
3	5	5	36.0	35.5	5	36.0	33.5
4	5	5	36.0	32.2	5	36.0	33.5
5	5	5	36.0	30.83	5	36.0	33.5
6	5	5	36.0	31.0	5	36.0	33.5
7	5	5	36.0	28.38	5	36.0	33.5
8	5	5	36.0	29.22	5	36.0	33.5
9	5	2	3.0	3.0	2	3.0	3.0
10	5	0	1.0	1.0	0	1.0	1.0
11	5	0	1.0	1.0	0	1.0	1.0
12	5	0	1.0	1.0	0	1.0	1.0
13	5	5	10.0	8.5	5	10.0	8.5
14	5	0	1.0	1.0	0	1.0	1.0
15	5	0	1.0	1.0	0	1.0	1.0
16	5	1	6.0	3.5	1	6.0	3.5
17	2	1	8.0	5.0	2	8.0	4.5
18	2	2	8.0	5.6	2	8.0	4.5
19	2	2	8.0	5.5	2	8.0	4.5
20	2	2	9.0	6.4	2	9.0	5.0
21	5	5	61.0	57.5	5	61.0	57.5
22	10	10	40.0	40.0	10	40.0	40.0

the average number of burned nodes is different for several instances. This is because alternative first-stage solutions have been found.

By comparing the runtimes in Table 3 it is clear that the decomposition approach is much more efficient in dealing with harder instances than the simpler implementation of the robust model, reducing the runtime from almost 2 hours to simply 3 minutes in instance 8. However, in simpler problems, such as instance 20, the decomposition approach is slower. This is due to the number of optimizations that are done in the decomposition model instead of the single optimization done in the static robust model, as indicated by the large runtime of the subproblem. Also, notice that instance 20 has a large number of scenarios. We can also observe that the number of iterations required by the decomposition approach is very small, ranging between 2 and 3.

Table 3 Performance metrics of the robust model

	Robust model	Decomposition approach			
Instance	Runtime (*seconds*)	Runtime (*seconds*)			Iterations
	Total	Total	Master	Sub	
1	16	44	27	17	3
2	266	139	99	40	3
3	718	149	98	50	3
4	1304	161	100	61	3
5	1131	170	99	70	3
6	2435	178	100	78	3
7	2159	189	98	90	3
8	6538	193	98	95	3
9	5	11	5	5	2
10	3	6	3	3	2
11	3	6	3	3	2
12	3	6	3	3	2
13	12	34	18	15	3
14	6	9	3	6	2
15	6	9	3	6	2
16	7	24	10	13	3
17	19	54	13	40	3
18	24	65	13	52	3
19	46	99	13	86	3
20	76	129	15	113	3
21	54	63	57	7	3
22	26	52	31	21	3

6 Conclusion and Future Work

We introduced a new robust model for the problem of prepositioning resources for fire suppression under uncertainty. The uncertainty is modelled by a set of scenarios representing fire ignitions locations, wind speed and direction. As the size of the model may become too large when the number of scenarios grow, we devised a row-and-column decomposition algorithm to attempt to solve the problem more efficiently. Computational experiments showed that for large instances, the decomposition approach clearly outperforms the direct one.

It is well known that solving a problem by taking into account in the objective only the worst-case scenario, as done in the current work, has the drawback of generating solutions that do not take into account what happens in the other scenarios. From this observation, two lines of future research could be considered. One is exploring alternative first-stage solutions that lead to a lower burned area on average. This can be done by solving the proposed model and then reoptimizing by imposing a new constraint ensuring that in the worst-case the total burned area cannot exceed a given value. Another direction is to explore distributionally robust optimization, which allows to combine the worst-case perspective with the expected value of the burned area.

Acknowledgements This work was supported by the Portuguese Foundation for Science and Technology (FCT) within project: UID/04561/2025 (CEMS.UL), the R&D Unit Project Scope UID/00319/Centro ALGORITMI (ALGORITMI/UM) and within project PCIF/GRF/0141/2019 "O3F—An Optimization Framework to Reduce Forest Fire", https://doi.org/10.54499/PCIF/GRF/0141/2019. This work was also supported by CIDMA under the Portuguese Foundation for Science and Technology (FCT, https://ror.org/00snfqn58) Multi-Annual Financing Program for R&D Units, grants UID/4106/2025 and UID/PRR/4106/2025.

References

1. Martell, D.L.: A review of recent forest and wildland fire management decision support systems research. Curr. For. Rep. **1**, 128–137 (2015)
2. Martell, D.L.: A review of operational research studies in forest fire management. Can. J. For. Res. **12**(2), 119–140 (1982)
3. Mendes, A.B., e Alvelos, F. P.: Iterated local search for the placement of wildland fire suppression resources. Eur. J. Oper. Res. **304**(3), 887–900 (2023)
4. Granda, B., León, J., Vitoriano, B., Hearne, J.: Decision support models and methodologies for fire suppression. Fire **6**, 2 (2023)
5. Hof, J., Omi, P.N., Bevers, M., Laven, R.D.: A timing-oriented approach to spatial allocation of fire management effort. For. Sci. **46**(3), 442–451 (2000)
6. Finney, M.A.: Fire growth using minimum travel time methods. Can. J. For. Res. **32**(8), 1420–1424 (2002)
7. Finney, M.A. et al.: An overview of flammap fire modeling capabilities. In Fuels Management—How to Measure Success: Conference Proceedings, vol. 28, p. 30. USDA Forest Service, Rocky Mountain Research Station, Fort Collins, CO (2006)
8. Wei, Y., Rideout, D.B., Hall, T.B.: Toward efficient management of large fires: a mixed integer programming model and two iterative approaches. For. Sci. **57**(5), 435–447 (2011)
9. Belval, E.J., Wei, Y., Bevers, M.: A mixed integer program to model spatial wildfire behavior and suppression placement decisions. Can. J. For. Res. **45**(4), 384–393 (2015)
10. Alvelos, F.: Mixed integer programming models for fire fighting. In International Conference on Computational Science and Its Applications, pp. 637–652. Springer (2018)
11. Harris, M.G., Forbes, M.A., Taimre, T.: Logic-based benders decomposition for wildfire suppression. Computers & Operations Research, p. 106392 (2023)
12. Birge, J.R., Louveaux, F.: Introduction to Stochastic Programming. Springer-Verlag (1997)
13. Ben-Tal, A., Nemirovski, A.: Robust convex optimization. Math. Oper. Res. **23**, 769–805 (1998)
14. Nemirovskiĭ, A.S., El Ghaoui, L., Ben-Tal, A.: Robust optimization. Princeton Series in Applied Mathematics. Princeton University Press (2009)
15. Erin J Belval, Yu Wei, and Michael Bevers. A stochastic mixed integer program to model spatial wildfire behavior and suppression placement decisions with uncertain weather. Can. J. For. Res. **46**(2), 234–248 (2016)
16. Mendes, A.B., Pereira e Alvelos, F.: A robust optimisation approach for the placement of wildland fire suppression resources. Available at SSRN 4385923 (2023)

17. Bienstock, D., Özbay, N.: Computing robust basestock levels. Discret. Optim. **5**(2), 389–414 (2008)
18. Agra, A., Christiansen, M., Figueiredo, R., Hvattum, L.M., Poss, M., Requejo, C.:The robust vehicle routing problem with time windows. Comput. Oper. Res. **40**(3), 856–866 (2013)
19. Zeng, B., Zhao, L.: Solving two-stage robust optimization problems using a column-and-constraint generation method. Oper. Res. Lett. **41**(5), 457–461 (2013)
20. Richard, R.C.: A mathematical model for predicting fire spread in wildland fuels, vol. 115. Intermountain Forest & Range Experiment Station, Forest Service, USA (1972)
21. Alvelos, F.: pyo3f—a python framework for wildfire-related optimization, part i: Design and fundamentals. In: Alvelos, F., Martins, I., Rocha, A. (eds.) Advances in Optimization and Wildfire: Proceedings of the First Optimization and Wildfire Conference, October 1–4, 2024, Luso, Portugal. Springer (2025)
22. Alvelos, F.: pyo3f—a python framework for wildfire-related optimization, part ii: Usage. In: Alvelos, F., Martins, I., Rocha, A. (eds.) Advances in Optimization and Wildfire: Proceedings of the First Optimization and Wildfire Conference, October 1–4, 2024, Luso, Portugal. Springer (2025)
23. Alvelos, F., Marto, M., Mendes, A.: Dispatching, positioning and routing resources for wildfire initial attack. Networks. In press (2025)
24. Andrews, P.L.: The rothermel surface fire spread model and associated developments: a comprehensive explanation. usda forest service, rocky mountain research station. Technical report, General Technical Report RMRSGTR-371.(Fort Collins, CO) (2018)

Wildfire Management: An Operational Research Perspective

David L. Martell

Abstract This paper is based on a talk delivered at the 2024 Optimization and Wildfire Conference in Luso, Portugal, on October 2, 2024, during which I characterized forest and wildland fire management from an operational research perspective. I present a brief description of fire management as it is practiced in the province of Ontario, Canada, and focus on how aircraft are used to detect and suppress wildfires there. I then describe how my colleagues and I are developing an aerial detection probability model and daily airtanker deployment models. I briefly describe what I consider to be three important current and emerging problems that I believe operational researchers can help resolve, and conclude by stressing the importance of close collaboration between researchers and wildfire managers when they do so.

Keywords Wildfire management · Operational research · Daily airtanker deployment · Queueing models · Aerial detection planning · Detection probability

1 Introduction

Forest and wildland fire management poses many interesting challenges and has been of interest to a small but growing community of operational researchers since Shepard and Jewell [20] first focussed their attention on wildfire management in the 1960s. They supervised a number of graduate students that studied wildfire management problems, one of whom was Parks [16] who published what I believe was the first peer-reviewed wildfire management paper in the OR literature—a model of the suppression of an initial attack fire.

When I completed what I then considered to be a comprehensive literature review in 1981, [11] I identified 200 peer-reviewed journal and technical grey literature wildfire-related OR papers that had been published during the years 1961 through 1981. Interest in the application of OR to wildfire management fire has grown in

D. L. Martell (✉)
Institute of Forestry and Conservation, John H. Daniels Faculty of Architecture, Landscape and Design, University of Toronto, Toronto, ON, Canada
e-mail: david.martell@utoronto.ca

F. Alvelos et al. (eds.), *Advances in Optimization and Wildfire*, Lecture Notes in Operations Research, https://doi.org/10.1007/978-3-032-03108-2_11

recent years because of increases in area burned and, more importantly, the increased frequency with which fire has impacted what wildfire managers refer to as the Wildland Urban Interface (WUI), "the area where human development meets or intermingles with the natural environment".[1] Increased burning of WUI areas has been attributed to a number of factors that vary from place to place including the growth of the WUI itself, changes in the composition of the vegetation (which wildfire specialists refer to as "fuel") and of course, climate change.

1.1 Overview

In this paper I describe some of the research in which my colleagues and I are currently engaged in the province of Ontario, one of the ten provinces and three territories in Canada. It is very much a description of our "work in progress" on two areas: aerial detection and daily airtanker deployment. My objective is not only to share some novel preliminary research results, but also to describe ***how*** we collaborate with Ontario's wildfire management organization and what I consider to be the importance of doing so.

I will begin by presenting a brief description of the province of Ontario, a summary of the extent to which fires occur in and burn Ontario's forests, and the province's fire management organization and policy. I will then describe the current status of our on-going research on daily airtanker deployment and Ontario's aerial detection system. While that sequence might strike some readers as being "backwards", the reason I chose to address those two topics in that order will soon be obvious. I will conclude by identifying what I consider to be three important current and emerging problems that believe operational researchers can help resolve, and some of the ways in which I believe they might best do so.

Before proceeding further, it is important to keep in mind that:

(a) Fire is a natural component of many forest and wildland ecosystems. It can and does have beneficial impacts on natural ecosystem processes, but it also can and often does have very significant detrimental social and economic impacts on the people that inhabit, live near, use and otherwise depend upon those ecosystems.
(b) The extent to which and how fire can and should be "managed" varies by ecosystem and there are no "one size fits all" policies and practices that can and should be applied to all flammable forest and wildland ecosystems.

[1] FireSmart Canada, https://firesmartcanada.ca/about-firesmart-2/the-wildland-urban-interface-wui/, accessed January 13, 2025.

2 Forest Fire Management in Ontario

Canadians consider Canada to be a forest nation and Ontario is very much a forested province. Ontario is covered with four forest regions; the Hudson Bay Lowlands, the boreal forest, the Great Lakes-St. Lawrence forest and the deciduous forest in the south.[2]

As of September 25, 2024, the 10-year average number of fires per year was 691 and the 10-year average area burned per year was 199,954 ha.[3] [21] reported that during the 1976–2001 period, 23% of the fires that were reported in Ontario were caused by lightning and these fires accounted for 83% of the area burned.

The Forest Fires Prevention Act (FFPA)[4] is the provincial government legislation that specifies the 975,981 km^2 Fire Region, the portion of the province in which Ontario's Ministry of Natural Resources (MNR) is responsible for wildfire management. Its 2014 Wildland Fire Management Strategy calls for every fire to receive an "appropriate response" that can vary from aggressive suppression to monitoring.

The Fire Region is partitioned into two regions (the Northeast and Northwest) that are further partitioned into a total of 13 sectors. Each region has a Regional Response Centre that is managed by a Response and Operations Manager who is responsible for managing the MNR's fire management program within that region.

Each day during the fire season, the staff of each Regional Response Centre includes a Fire Response Specialist that is referred to as the Regional Duty Officer (RDO). The duties of the RDOs include deciding upon the daily deployment and dispatch of airtankers. The MNR has a Provincial Response Centre that is managed by its General Manager Response and Operations whose duties include coordination of the two regional response centres.

2.1 *Use of Aircraft for Forest Fire Management in Ontario*

Ontario is a large, sparsely populated province that is covered by a sparse network of primary, secondary and tertiary roads. The area of the province of Ontario is 1.1 million km^2, of which 892 thousand km^2 is land and 159 thousand km^2 is water. Given the 2021 population of 14.2 million people, its population density is 15.9 people/km^2.

[17] used several different road databases to estimate road density in Ontario and other Canadian provinces. The Global Roads Inventory Project (GRIP) dataset revealed a road density in Ontario of only 0.07 km/km^2, which is much less than the

[2] Ontario's forest regions are described at https://www.ontario.ca/page/forest-regions, accessed February 3, 2025.

[3] Canadian Interagency Forest Fire Centre (CIFFC) National Fire Situation Report For: September 25, 2024. https://ciffc.net/situation/2024-09-25. Accessed February 3, 2025.

[4] Forest Fires Prevention Act, R.S.O. 1990, c. F.24. https://www.ontario.ca/laws/statute/90f24 accessed February 3, 2025.

GRIP road density that [15] reported for Europe (greater than 2.5 km/km^2) and the United States (greater than 1 km/km^2).[5]

Time is of the essence in wildfire management and although some fires, particularly human-caused fires that occur near heavily populated areas in Ontario, are readily accessible by road, aircraft are used extensively for fire management in Ontario for detection, initial attack and to support large fire suppression operations.

2.2 Airtanker Management

The province of Ontario has a fleet of 9 CL-415 airtankers and 3 DH-6 Twin Otter airtankers that are used primarily for initial attack and, to a very limited extent, to support suppression operations on large, escaped fires. Ontario sometimes temporarily lends some of its airtankers to other fire management agencies or supplements its fleet by temporarily borrowing airtankers from other agencies. [22] reported that 34% of the fires that were subjected to initial attack in Ontario during the years 2001–2019 received airtanker action.

Airtanker management takes place within a complex hierarchy of strategic, tactical and operational decision-making problems, some of which my colleagues and I have studied and continue to study. Airtanker management is hierarchical in the sense that many decisions taken in one level of the hierarchy are constrained by decisions taken in the higher levels of the hierarchy and in turn, constrain decision-making in the levels below them.

Strategic airtanker management calls for deciding how many of what kind of airtankers to purchase or lease and where to home base them. Tactical decision-making includes deciding where airtankers should be deployed or pre-positioned each day to minimize response times. Operational decision-making involves deciding what airtankers should be dispatched to what fires, how many drops should be delivered to each fire, where those drops should be placed, and if and when the airtankers should be pulled off a fire because they are proving to be ineffective or they are required on a higher priority fire.

Our approach to the decomposition problems posed by that hierarchical structure has been to develop subsystem models that account for such linkages. Home-basing decision-making is constrained by higher level purchasing and leasing decision-making. Daily airtanker deployment decision-making is influenced by higher level home-basing decision-making and in turn, constrains initial attack dispatch decision-making.

The Ontario initial attack study [12] was initiated in response to a Management Board of Cabinet directive that the Ontario Ministry of Natural Resources develop a rationale for how best to update its then aging air fleet. We developed a strategic

[5] It is, however, important to note that [17] found that the GRIP dataset includes only 13% of the primary, secondary and tertiary road network in Ontario.

simulation model of the use of airtankers, fire fighters and transport helicopters that was used to inform Ontario's purchase of 9 CL-215 airtankers.

[10] developed a mathematical programming model that was used to help identify a home-basing strategy for Ontario's airtanker fleet that would minimize the average annual cost of satisfying daily airtanker deployment demands they assumed would be used each day of the fire season, that were elicited from a group of Ontario's fire management personnel.

My colleagues and I have studied the daily deployment of airtankers for initial attack for many years and we continue to do so. Ontario has many primary airports or bases at which they can deploy their airtankers each day, and each day their staff must decide what airtankers should be deployed at what base and their alert status—that is, how quickly they must be airborne once they are dispatched to a fire.

Since new initial attack fires are reported or "arrive" at rates that vary throughout the day, deciding how many airtankers to deploy at each airtanker base calls for the design and control of a complex queueing system with several service centres (airports), each of which has a designated number of servers (airtankers) at which customers (fires) "arrive" at rates that vary throughout the day. That fires arrive at rates that vary throughout the day, that some require airtanker action more than others and that the servers must travel to and from (and sometimes between) fires complicate daily airtanker deployment decision-making.

Some of our attempts to solve variants of the daily airtanker deployment problem include [7, 13] and [8]. [13] extended [3] daily initial attack transport helicopter deployment model to develop an airtanker deployment model they field tested in Ontario. [7] developed a simulation model that captured the interaction of bases but when they did so, they used very rough estimates of the service time parameters because that is all they had available. [8] developed a time-dependant variant of the [9] hypercube queueing model.

We recognized that our models were increasingly realistic but not sufficiently realistic, largely because we lacked good onsite or "over the fire" airtanker performance data that we realized was needed to parameterize them. We therefore decided to continue to study the daily airtanker deployment problem but we did not attempt to convince MNR staff with whom we continued to collaborate, to use our models to inform their decision-making.

My colleagues and I could formulate and "solve" models of such decision-making problems, but when we did so we are challenged by the fact that it is very difficult to acquire the airtanker operations data required to model what actually happens at a fire such as the number of drops made, the distance between the fire and the lake from which the airtankers scoop water and the time required to do so. The cost of hiring research assistants, assigning helicopters to take them to and hover near active airtanker operations to collect such data would not only pose "over the fire" aircraft safety challenges, but would be prohibitively expensive.

In 2009 a senior manger of Ontario's fire program with whom I had very much earlier collaborated on the development and field testing of a relatively simple daily airtanker deployment model [13] emailed me a map (see Fig. 1 in [4]) that displayed the location, every 30 s, of a set Ontario's airtankers that had been engaged suppression operations earlier that day.

His email was brief—he simply wrote, to paraphrase,—I thought you might be interested in what our airtankers were doing today. His primary objective was not really to describe what the airtankers had been doing that day but rather, to draw my attention, to the existence of data (time-stepped GPS data which we describe as heartbeat data) of which I had hitherto been unaware, that he knew would be of interest to me and my students. I quickly requested and received a comprehensive set of Ontario's airtanker heartbeat data. One of my students then used that to model airtanker operations in far more detail than had ever been possible in the past as described in [4]. ***I would not have learned about that data and its potential use had I not had a long very close collaborative research partnership with Ontario's wildfire management program.***

We have since focussed our attention on the development of a more realistic open source simulation model. We began by developing a SimPy model in which we incorporated a very simple—dare I say simplistic—airtanker dispatch model and a suppression model. Fortunately, [22] recently developed an airtanker dispatch model and our hope and expectation is that we can convince the MNR to field test it during the 2025 fire season. ***Their willingness to do so will in no small part be influenced by the fact that we have collaborated closely with them for so many years.***

3 Ontario's Forest Fire Detection System

My colleagues and I have also been studying Ontario's forest fire detection system for a number of years. The Ontario Department of Lands and Forests (ODLF), the provincial government department that preceded the MNR, began building wooden fire lookout towers early in the twentieth century. It shifted to 90-foot and 100-foot steel towers in the 1920's and [18] reported that Ontario had 329 steel lookout towers in 1960.

In the late 1960s the ODLF grew increasingly concerned about the cost-effectiveness of towers and initiated several studies (e.g., [5]) to document their performance. Those reports and recognition that the public detected and reported many fires contributed to the ODLF's decision to transition from fixed lookout towers to aerial detection patrols in the 1970's and the MNR now relies on the public and aerial detection patrols to detect and report wildfires.[6]

The management of the MNR's aerial detection system poses a number of challenging decision-making problems that are amenable to OR approaches, one of which

[6] The MNR sometimes enhances its detection system by using what it describes as "loaded patrols" during very high hazard conditions. A helicopter carrying an initial attack crew and its equipment will fly a specified detection patrol route. If and when they detect a fire, the helicopter will land and drop the initial attack crew and its equipment off near the newly detected fire. Although the use of loaded patrols is costly, the very significant reduction in the time between when a fire is detected and when initial attack begins is thought to outweigh that cost.

is daily aerial detection patrol route planning. Each year the MNR charters a small fleet of fixed-wing aircraft for detection purposes for most of the fire season and the crew of a detection patrol aircraft includes the pilot and a detection observer. Although the pilot's primary role is to fly the aircraft, both the pilot and the detection observer search for fires.[7] Each day the detection planner, in consultation with the two Regional Duty Officers, must decide when and where they will fly patrols to search for fires.

The MNR's detection objective is not to find all fires as quickly or at as small a size as possible. Recall that the MNR's Strategy calls for every fire that occurs in Ontario to receive an ***appropriate*** response. Natural lightning-caused fires burning in remote areas that are having beneficial impacts on the ecosystems in which they are burning do not "need" to be detected and reported as quickly as fires that are burning in flammable vegetation near communities and other values at risk when the fire hazard is high or extreme. There is also little need to fly detection patrols along busy highways that are driven by people with mobile phones who will report any fires they see, or close to communities inhabited by people who will report such fires.

Detection patrol aircraft and their crew are costly but effective resources, and the detection planner must decide how best to deploy them each day. To borrow from the supply chain logistics paradigm, one might view the detection system objective as using the ***right*** resource (i.e., detection patrol aircraft or the public) to find the ***right*** fire at the ***right*** time at the ***right*** cost. Fortunately, the MNR developed the Aerial Detection Demand Index (ADDI) that addresses the last three factors.

The ADDI [14] is a spatially explicit measure of the need for an aerial detection aircraft to visit each of 2,571 20 km $\times$ 20 km cells in Ontario's fire region each day. It accounts for several factors including the probability that there are undetected fires burning in the cell, the values at risk in that cell and the likelihood that any undetected fires will be detected and reported by the public.

Consider the following simple detection patrol route planning problem. Suppose the MNR has one detection patrol aircraft and its crew located at a particular airport, and the Regional Duty Officer wants that aircraft to fly a three hour patrol that will start and end at that airport. Given those constraints, the detection planner should specify a patrol that will pass through and near the "best" set of cells, a variant of Balas' [2] prize collecting problem. Unfortunately, the detection patrol routing problem is more challenging than the prize collecting problem because the search process is stochastic—if a detection patrol flies through a 20 km $\times$ 20 km cell that contains undetected fires, the pilot or observer may or may not detect any or all of those fires. It is even less likely that they will detect all the undetected fires at greater distances from the flight line.

The MNR drew on the search and rescue literature and assumed that a detection pilot and observer could "effectively" scan up to 20 km from the flight line. Each day the RDO would decide whether or not aerial detection was required and if so,

[7] The detection patrol flight crew does not always include a detection observer. From time to time the MNR forgoes the hiring of detection observers relies upon the pilot to detect and report fires. It sometimes assigns other MNR employees to accompany the pilot on high hazard days.

the parts of each region that should be "looked at" and how often (i.e., number of patrols) they should be looked at. The detection planner would then lay out a set of patrols and present them to the RDO who would approve or suggest how they might be revised.

My colleagues and I believed that if we had a detection probability model which related the probability that a fire would be detected to the distance from the flight line to the undetected fire and other factors, such detection probabilities could inform and enhance the daily detection patrol route planning. Unfortunately, to our knowledge, no such detection probability model exists. Fortunately, my colleagues and I were able to develop a detection probability function that we believe will address that need.

3.1 The Need for an Aerial Detection Probability Model

Rego and Catry [19] used distance sampling methods to develop a model that relates the probability that a fire will be detected by a lookout observer in a fixed tower, to the distance of the fire from the tower. Hower, persistent or repeated scanning by a human (or a camera) in a stationary tower, of the area near that tower, differs dramatically from short term scanning of portions of the landscape from a moving aircraft.

We "know" that the probability that an observer will detect a fire depends on the size and colour of the smoke column produced by the fire, the distance of the fire from the flight line and many other factors including atmospheric visibility and the location of the sun with respect to the line between the aircraft and the undetected fire. Unfortunately, we don't know the parameters of that detection probability function.

One could, in principle, design and carry out controlled experiments to develop an aerial detection probability model using smoke generators and or small experimental fires to study the effectiveness of fire lookout tower observers. The cost of carrying out such experiments in both time and money are such that it was never seriously considered. Fortunately, we realized that Ontario's operational fire data could be used to develop such a model at a very reasonable cost.

Recall that the MNR's airtankers are equipped with GPS heartbeat transponders. The MNR staff with whom we were collaborating on our study of the MNR detection system pointed out that their aerial detection aircraft were also equipped with GPS heartbeat transponders. We realized that the methods that Clark and Martell [4] had used to track airtankers could be used to track aerial detection aircraft, and we were able to determine precisely when and where they flew! That unfortunately, was not enough.

The MNR fire reports indicate where each fire that was detected by an aerial detection pilot or observer, MNR staff involved in the suppression of other fires or the public, was estimated to have been ignited. The GPS data tells us how close the patrol flew near to fires they detected. Unfortunately, it does not tell us when and how close it flew to fires that it missed because we do not know ***when*** every fire was ignited.

The MNR fire reports include the estimated clock time and date each fire was known or estimated to have been ignited. Unfortunately, we felt that data was not accurate enough to use to determine if and when a detection patrol flew near the fire. Suppose a fire started on day 1, a patrol flew near it on days 1, 2, 3 and 4, and it was detected on day 5. If we knew the time it was ignited on day 1 and that time was before the first flight passed by, we would have 4 unsuccessful looks and 1 successful look at the fire that we could use to develop a detection probability model in which the distance between the fire and the detection patrol flight line was an independent variable. Fortunately, because we had been collaborating with the MNR for many years, we were aware that we could use an objective method for estimating when some (lightning-caused) fires were ignited.

Wotton and Martell [21] describe the development of a model that can be used to predict the probability that a lightning strike will ignite a fire in Ontario. Modern lightning strike detection systems report the time and location of lightning strikes within designated radii of the sensor. The timing is precise, but the reported location is somewhat uncertain. Fortunately, Wotton, who has collaborated with the MNR for many years, has since continued his study of lightning fire ignition processes and lightning strike sensors and was able to develop an algorithm to estimate which lightning strike (and therefore also the time of the strike) had ignited most of the lightning fires that occurred in Ontario during our study period. We therefore had good estimates of the time each lightning fire was ignited, so we could identify when and how many unsuccessful looks at each lightning fire took place. We were then able to use that data to develop a logistic regression model that relates the detection probability to the distance from the flight line to the fire and other factors.

Our next step will be to develop a variant of [2] prize collecting problem to develop a detection patrol route evaluation algorithm to inform the detection patrol route decision-making. ***We could not have developed the detection probability model and we would not be in a position to develop a detection patrol routing model had we not been collaborating closely with the MNR for many years.***

4 Some Current and Emerging Canadian Wildfire Management Challenges

The "forest fire problem" was, until recently, viewed by most Canadians as one that primarily impacted people that lived and worked in and near forested areas. That is certainly no longer the case for a number of reasons including climate change, how our forests are and have been managed, and how we use our forests. Wildfires that threatened and burned through and destroyed communities, and fires that emitted smoke that severely impacted air quality in far distant towns and cities were, for the most part, considered to be rare events. Many Canadians have now unfortunately come to expect and fear that they might be threatened by wildfires every year, and that they need to prepare to deal with such incidents if and when they arise.

Canadian wildfire management is no longer "business as usual". Wildfire managers that once had the relatively simple tasks of preventing fires from occurring, detecting, and quickly suppressing those that did occur and limiting the growth of large fires that posed threats to communities and other values, now have to contend with many new challenges. Three of these are (1) a lack of resources, (2) a need to treat fuels to protect communities and other values at risk and (3) deal with smoke and carbon emissions.

4.1 Wildfire Management Agency Staffing and International Resource Sharing

Canadian wildfire management is primarily a provincial and territorial responsibility and Canadian wildfire managers have long recognized the importance of resource sharing. The Canadian Interagency Forest Fire Centre (CIFFC) is a not-for-profit corporation that was established in 1982. One of its objectives is to facilitate the sharing of fire management personnel and equipment including aircraft, both within Canada and with other countries, and it has fulfilled that objective very well.

The 2023 fire season was a record season and CIFFC helped Canadian wildfire management agencies borrow more than 5,000 people from other countries. That level of borrowing is not expected to become the norm in the near future, but the number of fire management personnel that Canadian wildfire management agencies have borrowed from other countries in recent years suggests that Canada does not have enough experienced fire management personnel. That there is a need to study fire management resource needs and how best to manage them under the high degree of uncertainty that is characteristic of wildfire management is obvious. Operational researchers would be quick to recognize that they could ***help*** resolve such problems.

4.2 Landscape Level Fuel Management and Community Protection

Communities located in or near flammable forests are sometimes, and in recent years, increasingly, threatened and destroyed by wildfires. Notable examples in Canada in recent years include the 2003 Okanagan Mountain Park and McLure fires in British Columbia, the 2016 Fort McMurray fire in Alberta, the 2021 Lytton fire in British Columbia and most recently, the 2024 Jasper fire in Alberta. That there is a need for good prevention, detection, initial attack, and large fire management systems and to FireSmart homes and other structures in communities that lie in the path of large intense fires is obvious. There is also growing recognition of a need to manage fuels, for example to create fuel breaks or treat fuels (e.g., create fuel breaks or replace highly flammable vegetation with less flammable vegetation) to reduce the likelihood that a fire that approaches a community will ignite homes and other structures within it.

Fuel management is very costly and poses difficult questions such as when, where and how it should take place—the width of a fuel break being a simple example for which there are no simple answers. Some observers have suggested there is a need to look well beyond communities and infrastructure and treat some fuels on the larger landscape in which they are embedded. Landscape level fuel treatment programs would be very costly. [6] suggested that forest companies could perhaps schedule the timing and location of their roads and harvesting operations in ways that might strategically fragment the landscape such that fire growth potential is reduced and their allowable cuts increase to counter the increased costs of doing so. [1] developed an optimization model and applied it to a 20,790 portion of a forest management unit in the province of Alberta. They made some sweeping assumptions about the impact of harvesting on fire growth but found that in theory, they could FireSmart the management of their study area and increase the annual harvest volume by 8.1%.

Such results beg the question—to what extent can and should the extensive landscape be fragmented to protect the communities and other values embedded within it? That has led to many attempts to develop spatially explicit landscape management plans designed to mitigate fire losses. Forested landscapes are complex ecosystems that are used for many purposes, are home to many species, and are inhabited by people that value, enjoy and use them in many diverse ways. FireSmarting of extensive flammable forest landscapes may well prove to be advisable, but should be informed by the expertise of forest managers, ecologists, economists, other social scientists and of course, the residents of communities embedded in those landscapes. I believe that operational researchers that can formulate and solve spatially explicit stochastic models will have an important role to play.

4.3 Smoke and Carbon Emissions

Smoke and carbon emissions have recently emerged as urgent concerns that need to be addressed. Canadians that live in northern forested communities have long "put up" with the health threats posed by smoke, but the transport of smoke from Canadian wildfires to large urban centres in southern Canada and the United States has drawn attention to the health threats of wildfire smoke.

I expect that international recognition of the level of such emissions will soon lead to demands that Canadians reduce them. Unfortunately, that is a difficult problem for which there appear to be no simple answers.

One could of course, attempt to suppress all fires, but we lack the resources to do so and in the long term, that would build up a carbon "time bomb" that we would pass on to future generations of Canadians. Another possibility is to use the forest as a source of biofuels to replace our use of fossil fuels, but that technology is currently in its infancy.

That Canada contains massive peat deposits that burn and emit massive amounts of carbon when dry, and will burn even more in the future due to climate change, poses an even more difficult challenge. Our wildfire suppression systems are designed to prevent and contain fires that burn surface and crown fuels and the deep burning organic soil on the perimeters of large fires. The interior of large fires are left to burn until they run out of fuel or rainfall extinguishes them. In the past, we could rely on winter snowfall to melt and extinguish burning peat deposits that burned through the winter under the snow, when it melted in the spring. Unfortunately, recent experience indicates we can no longer depend on winter snowfall to extinguish all "over-wintering" wildfires in the spring.

Canadians very clearly have to "do something" about our growing wildfire smoke and carbon emissions for which there are no obvious answers. We will need to call on health care professionals, occupational health and safety professionals, ecologists, forest managers, economists and other social scientists, to develop feasible solutions. I expect operational researchers could play a role in their implementation.

5 Some Suggestions for Those that Want to Apply OR to Wildfire Management Problems

I chose to focus this paper on two of my on-going research projects, but there remain many other important open problems that I have not discussed. Climate change, land management practices and other factors have brought us to a point where wildfire disasters are now the norm—the January 2025 Los Angeles fires that burned as I wrote this paper being a case in point. I am pleased but not surprised that such disasters have led to an increase, dare I say an explosion, in the number of operational researchers who want to bring their knowledge and expertise to bear on wildfire management problems. Faced with what is unfolding, some might expect me to provide a long list of suggestions to operational researchers that wish to join the battle. I could of course, draw on my own experience and draw up such a list, but I am well aware that what "worked" and did not work for me and my colleagues in Ontario is not likely to address all the needs of operational researchers elsewhere. That being said, I will close with two suggestions that I expect will serve well those operational researchers that wish to help solve real wildfire management problems;

(1) Keep one foot firmly established in your OR research community so you can keep abreast of advances in OR, and the other, in a wildfire agency or agencies with which you collaborate, and listen carefully to their staff when they identify what they consider to be important planning and decision-making problems.
(2) Pay close attention when they critique your proposed solutions.

Acknowledgements I have been fortunate throughout my career, to have been given many opportunities to collaborate with many wildfire professionals that were employed by a wildfire management organization that was and remains very committed to supporting research, and to adopting findings that showed promise, in collaboration with many students and research assistants that shared my vision and chose to work with me.

The research described in this paper was supported by the Natural Sciences and Engineering Research Council of Canada (NSERC) and the Ontario Ministry of Natural Resources (MNR). It was carried out in collaboration with many current and former MNR fire management program staff including T. Ball, G.E. Doan, R.J. Drysdale, G. Harland, A. Johnson, D. Leonard, M. MacQuarrie, D. Manol, R. McAlpine, C. McFayden, P. Parker, S. Reany-Iskra, A. Tithecott and many former students and research assistants including M. Ansell, D. Boychuk, C. Buttazzoni, N. Clark, K. Islam, J. McLellan, Z. Turner-Debs, R. Santiago and B.M. Wotton.

References

1. Acuna, M.A., Palma, C.D., Cui, W., Martell, D.L., Weintraub, A.: Integrated spatial fire and forest management planning. Can. J. For. Res. **40**(12), 2370–2383 (2010). https://doi.org/10.1139/X10-151
2. Balas, E.: The prize collecting traveling salesman problem. Networks **19**(6), 621–636 (1989)
3. Bookbinder, J.H., Martell, D.L.: Time-dependent queuing approach to helicopter allocation for forest fire initial attack. Infor **17**(1), 58–70 (1979)
4. Clark, N.A., Martell, D.L.: The use of aircraft tracking GPS data to develop models of the use of airtankers in forest fire management. INFOR: Inf. Syst. Oper. Res. **57**:4, 535–562 (2020). https://doi.org/10.1080/03155986.2019.1706398
5. Drysdale, J.: Detection assessment and improvement studies, 1966–1968. Kenora, Oct. 28, 1968. Ontario department of lands and forests AFMX/LIB 190 F. (1968)
6. Hirsch, K., Kafka, V., Tymstra, C., McAlpine, R., Hawkes, B., Stegehuis, H., Quintilio, S., Gauthier, S., Peck, K.: Firesmart forest management: a pragmatic approach to sustainable forest management in fire-dominated ecosystems. For. Chron. **77**, 357–363 (2001)
7. Islam, K.M.S., Martell, D.L.: Performance of initial attack airtanker systems with interacting bases and variable initial attack ranges. Can. J. Forest Res. **28**(10), 1448–1455 (1998)
8. Islam, K.S., Martell, D.L., Posner, M.J.: A time-dependent spatial queueing model for the daily deployment of airtankers for forest fire control. INFOR: Inf. Syst. Oper. Res. **47**(4), 319–333 (2009)
9. Larson, R.C.: A hypercube queueing model for facility location and redistricting in urban emergency services. Comput. Oper. Res. **1**, 67–95 (1974)
10. MacLellan, J.I., Martell, D.L.: Basing airtankers for forest fire control in Ontario. Oper. Res. **44**(5), 677–686 (1996)
11. Martell, D.L.: A review of operational research studies in forest fire management. Can. J. Forest Res. **12**(2), 119–140 (1982)
12. Martell, D.L., Drysdale, R.J., Doan, G.E., Boychuk, D.: An evaluation of forest fire initial attack resources. Interfaces **14**(5), 20–32 (1984)
13. Martell, D.L. Tithecott, A.: Development of daily airtanker deployment models. In: Proceedings of the 1991 Symposium on Systems Analysis in Forest Resources. Charleston, S.C. Compiled by M.A. Buford 3–6 March 1991
14. McFayden, C.B., Woolford, D.G., Stacey, A., Boychuk, D., Johnston, J.M., Wheatley, M.J., Martell, D.L.: Risk assessment for wildland fire aerial detection patrol route planning in Ontario, Canada. Int. J. Wildland Fire **29**(1), 28–41 (2020)
15. Meijer, J.R., Huijbregts, M.A., Schotten, K.C., Schipper, A.M.: Global patterns of current and future road infrastructure. Environ. Res. Lett. **13**(6), 064006 (2018). https://doi.org/10.1088/1748-9326/aabd42

16. Parks, G.M.: Development and application of a model for suppression of forest fires. Manage. Sci. **10**(4), 760–766 (1964)
17. Poley, L.G., Schuster, R., Smith, W., Ray, J.C.: Identifying differences in roadless areas in Canada based on global, national, and regional road datasets. Conserv. Sci. Pract. (2022). https://doi.org/10.1111/csp2.12656
18. Rajala, R.A.: Feds, forests, and fire: a century of Canadian forestry innovation. xi, 116 p.: ill. Canada Science and Technology Museum. Catalogue number NM97–2/13E-PDF. (2005). https://publications.gc.ca/site/eng/9.860363/publication.html
19. Rego, F.C., Catry, F.X.: Modelling the effects of distance on the probability of fire detection from lookouts. Int. J. Wildland Fire **15**(2), 197–202 (2006)
20. Shephard, R.W., Jewell, W.S.: Operations research in forest fire problems. In A study of fire problems. National Academy of Sciences (USA). National Research Council Comm. Fire. pp. 145–165 (1961)
21. Wotton, B.M., Martell, D.L.: A lightning fire occurrence model for Ontario. Can. J. For. Res. **35**(6), 1389–1401 (2005)
22. Wheatley, M., Wotton, B.M., Woolford, D.G., Martell, D.L., Johnston, J.M.: Modelling decisions concerning the dispatch of airtankers for initial attack on forest fires in Ontario, Canada. Can. J. For. Res. **53**, 217–233 (2023)

Optimization and Wildfire: Supporting Strategic Mitigation of Extreme Events

Alan T. Murray

Abstract California has experienced significant increases in scale, frequency and intensity of wildfires, posing unprecedented threats to communities across the state. As a result, it has become vital to develop analytical tools, risk reduction strategies and operational plans that enhance wildfire resilience and co-existence. One approach is the use of vegetation treatment, which may involve prescribed burns, strategic fuel breaks, mastication and other interventions. A challenge is selecting the most impactful areas for such treatment, where viable projects must generally be contiguous and manageable in size. Spatial optimization modeling has much potential for addressing this selection problem, enabling the formalization of the planning process as well as offering approaches for identifying the best solutions possible. However, challenges remain as the underlying geographical detail and spatio-temporal characteristics are formidable. This paper highlights spatial optimization and geographic analytics to support vegetation treatment planning. Application results demonstrate the importance and challenges for optimization to address project area identification.

Keywords Co-existence · Patch contiguity · Spatial analytics

1 Introduction

While society faces a range of hazards on a daily basis, wildfire is an ever present threat to people and property across North America (and beyond), and is challenging communities in a variety of ways. On the one hand, fire is an important component of natural landscape maintenance and regenerative processes, but on the other, fires have become more extreme and devastating due to climate change, population growth, urban expansion, land use modification and decades of human fire intervention. The result has been historic levels of damage and destruction on regular basis when wildfires do occur. Balch et al. [1] note that the most deadly and destructive wildfires

A. T. Murray (✉)
Wildfire Resilience Initiative, University of California at Santa Barbara, Santa Barbara, CA, USA
e-mail: amurray@ucsb.edu

F. Alvelos et al. (eds.), *Advances in Optimization and Wildfire*, Lecture Notes in Operations Research, https://doi.org/10.1007/978-3-032-03108-2_12

in US history have been witnessed in recent years. Consider the situation in California summarized in Table 1. As an example, 2024 saw 2148 structures impacted (1716 destroyed and 432 damaged) along with disruptions to economic productivity and the loss of critical habitat, substantiating that the impacts of wildfire are not trivial, yet historic occurrences and costs are becoming the norm in fire prone regions where extreme conditions are no longer unusual. Consider the following fires in California:

- 2025 Palisades & Eaton (Los Angeles County)
- 2020 North Complex (Butte and Plumas Counties)
- 2018 Woolsey (Ventura and Los Angeles Counties)
- 2018 Camp (Butte County)
- 2017 Thomas (Ventura and Santa Barbara Counties)
- 2017 Tubbs/LNU Complex (Napa and Sonoma Counties)

These all occurred in the past 10 years, and are among the most costly ever observed in the US, with each in the billions of dollars. The significant rise in losses is shocking and unsustainable. Of the 8024 wildfire incidents in 2024 (Table 1), the Park (Butte and Tehama Counties north of Chico, 429,603 acres, 763 structures), Mountain (Ventura County north of Camarillo, 19,904 acres, 370 structures) and Franklin (Los Angeles County in Malibu, 4037 acres, 48 structures) fires are further examples of the significance of these devastating events driven by extreme conditions.

Many have concluded that it is critical to transition landscapes to more fire resilient conditions [2–5]. Doing so, however, is not a trivial task. It will be costly to achieve conditions more representative of historic fire adaptive ecosystems given constraints on safety and the protection of property that must be observed, but also will require considerable oversight guided by deliberate actions. This ultimately means strategic investment and planning will be fundamentally important in the realization of wildfire coexistence strategies, especially in and around the wildland urban interface (or WUI).

As wildfire resilience and coexistence will not be achieved by chance, the purposeful actions needed will be many. Strategies such as the creation and extension of defensible space, debris removal, decreasing vegetation densities, prescribed burns, land use conversion and others are among the primary courses of action required (see Parkins et al. [6]). Significant attention has focused on fuels treatment given the potential to reduce fire intensity and severity as well as limit spread and enhance suppression capabilities [7, 8], yet it is impossible to do this everywhere. As a

Table 1 Wildfire in California†

Year	Incidents	Acres	Fatalities	Structures
2024	8024	1,050,012	1	2148
2023	7127	324,917	4	71
2022	7477	331,360	9	876
2021	7396	2,569,386	3	3846

†CAL FIRE, https://www.fire.ca.gov/incidents, accessed 3/24/25

result, a number of spatial optimization based efforts have been developed to support strategic decision making that reduces wildfire risk and vulnerability, including the work of [9–14]. Qi and Zhuang [15] offer a review of optimization work along these lines. Recognition of competing considerations in fuels treatment prioritization has led to extensions of various sorts by [16–19], among others.

This paper reviews contemporary efforts to formalize strategic wildfire risk and vulnerability mitigation. In doing so the intent is to demonstrate the bridging taking place between wildfire management and spatial optimization advancements. The next section details an optimization approach specifically designed to support multi-objective decision making. Important features and properties of this approach are then reviewed. Next, major challenges in optimization capabilities are discussed. The paper ends with concluding thoughts.

2 Supporting Analytics

Calkin et al. [7] highlight that optimization is critical in addressing the growing threat of extreme wildfire, noting that doing so requires rather precise goals and objectives in mitigation efforts. Protecting against the loss of property and lives is at the top of the list, but critical habitat, threatened species, economic productivity, socio-economic implications, public health, etc. are also important concerns. Recognizing that strategic mitigation through fuel load reductions is essential, Ager et al. [9] suggest basic issues that modeling must support associated with treatment area identification and prioritization, including contending with limited resources/budgets, project management and spatial contiguity. The intention is to identify an individual patch (or project area) for fuels treatment, similar to the selected units shown in Figure 1. Murray et al. [20] observe that the approach of [9] is a type of land acquisition problem. In making this connection, Murray et al. [20] demonstrate that a spatial optimization model addressing the basic needs of treatment/project area (or patch) planning can be explicitly formulated.

Consider the following notation:

i = index of land units (entire set I)

k = index of objective attributes (entire set K)

w_k = importance weight for objective k

β_{ik} = benefit of mitigation/treatment in unit i for objective attribute k

α_i = area of unit i

N_i = set of units that neighbor unit i

p = number of patches (or project areas) to be treated

τ = maximum size of treated patch

$$X_i = \begin{cases} 1 \text{ if unit } i \text{ is selected for mitigation/treatment} \\ 0 \text{ otherwise} \end{cases}$$

$$V_i = \begin{cases} 1 \text{ if unit } i \text{ is selected patch sink} \\ 0 \text{ otherwise} \end{cases}$$

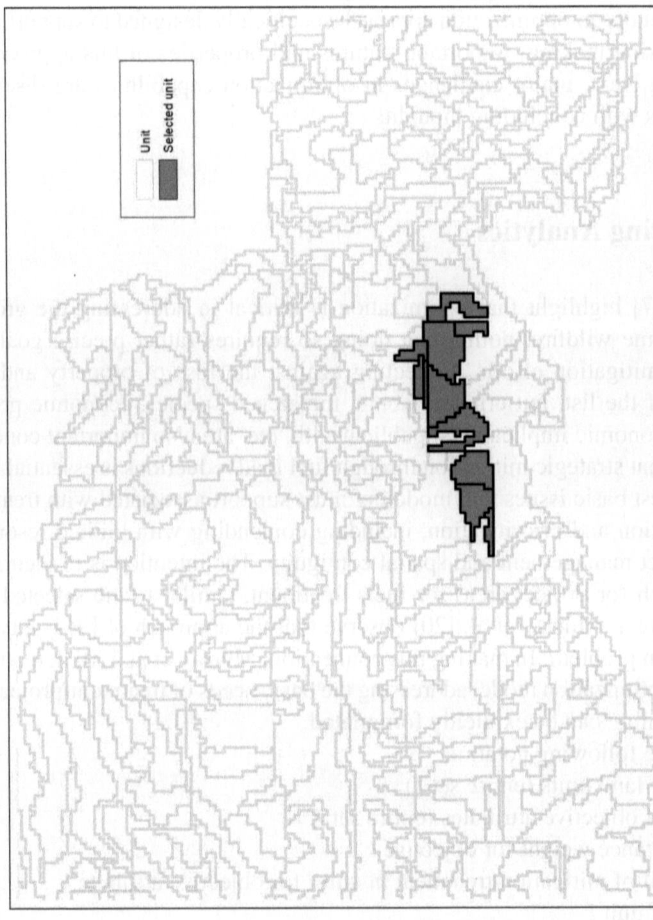

Fig. 1 Illustration of a single contiguous patch (Stanislaus National Forest)

Y_{ij} = accumulated flow from unit i to unit j destined for patch sink

Initial work formalizing patch selection in multi-objective terms in the context of wildfire risk mitigation is that of [16]. A weighting approach was used to integrate the various objectives in order to identify a single patch. Given the above notation, the basic model is approximately stated as follows:

$$\text{Maximize} \sum_{i \in I} \sum_{k \in K} w_k \beta_{ik} X_i \tag{1}$$

$$\text{Subject to} \sum_{i \in I} \alpha_i X_i \leq \tau \quad \forall i \in I \tag{2}$$

$$\text{Contiguity constraints} \tag{3}$$

$$X_i \in \{0, 1\} \quad \forall i \in I \tag{4}$$

The objective, (1), is to maximize the weighted combination of attributes associated with treatment selection of each unit. Constraint (2) stipulates that the patch (or project area) must not exceed a maximum size τ. Integer restrictions are imposed in constraints (4). Described in Ager et al. [16] but not specified mathematically was the need for spatial contiguity (3), of selected units for treatment. Worth noting is that a patch represents an administrative collection of units, for which a mitigation plan is to be executed. Given this management intent, there is nothing that prohibits two or more selected patches from neighboring each other.

Subsequent work by Pludow et al. [19] demonstrated that exact formulation of a multi-patch, multi-objective extension was possible. Without loss of generality, assume that $|K| = 2$ for illustrative purposes, where $k = 1$ represents wildfire resilience benefit of mitigation/treatment in unit i and $k = 2$ represents social vulnerability of unit i, as considered in [21]. The associated spatial optimization model to support the identification of the best patches (or project areas) for treatment can be structured as follows:

$$\text{Maximize} \sum_{i \in I} \beta_{i1} X_i \tag{5}$$

$$\text{Minimize} \sum_{i \in I} \beta_{i2} X_i \tag{6}$$

$$\text{Subject to} \sum_{j \in N_i} Y_{ij} - \sum_{j \in N_i} Y_{ji} \geq \alpha_i X_i - \tau V_i \quad \forall i \in I \tag{7}$$

$$\sum_{j \in N_i} Y_{ij} \leq \tau X_i \quad \forall i \in I \tag{8}$$

$$\sum_{i \in I} V_i = p \tag{9}$$

$$V_i \leq X_i \quad \forall i \in I \tag{10}$$

$$X_i \in \{0, 1\} \quad \forall i \in I \tag{11}$$

$$V_i \in \{0, 1\} \quad \forall i \in I \tag{12}$$

$$Y_{ij} \geq 0 \quad \forall i \in I, j \in N_i \tag{13}$$

Objective (5) indicates the intent to maximize attribute $k = 1$, or wildfire resilience in this case. Objective (6) specifies that attribute $k = 2$ is to be simultaneously optimized (social vulnerability in this case). Constraints (7) structure flow from selected units that must be routed to a patch sink and must conform to patch average size limits. If patches (or project areas) do not neighbor each other, than the size limit will be strictly enforced. However, when one or more patches neighbor each other, flow presenting unit size could be split (or shared) and would make the constraining condition(s) a limit on average size. Constraints (8) restrict flow out of a selected unit to be no more than the average bound. Constraints (9) establish the number of treatment patches to be identified. Constraints (10) stipulate that a patch sink can only exist among the selected patch units. Binary and non-negativity conditions are imposed in constraints (11–13).

3 Properties, Capabilities and Insights

As noted in Ager [22], (1–4) is a fundamentally important problem in fuels management, landscape restoration, risk reduction, habitat conservation, etc. Among many capabilities, developed planning support environments like ForSys, Land Tender and Planscape include solution of (1–4) using heuristics. Ager [22] indicates that developed heuristic approaches are able to impose contiguity within the patch/project area construction process. An additional feature is that different capabilities may be employed to identify patches, including consideration of secondary spatial features like compactness.

Seeking to more explicitly formalize (1–4), Murray et al. [20] derived a model formulation that could be solved using commercial mixed-integer programming software in order to ensure contiguity within a patch while identifying multiple patches simultaneously. An alternative formulation is suggested in Murray and Church [23] that has tighter model structure and is invariant in problem size as the number of patches changes, offering superior solution properties. The above formulation, (5–13), is a multi-objective extension of [23] detailed in [21], designed to enable trade-offs between many considerations to be simultaneously taken into account, with integrated hazard and social vulnerability addressed in this case.

At the heart of both (1–4) and (5–13) is ensuring that a patch (or project area) is not fragmented. Patch contiguity is a geographic property where selected units form a mutually connected area, where one can travel/move from any selected unit through other selected neighbors in order to get to any other selected unit. This can be observed for the 11 selected units forming a patch in Figure 1, with unit adjacency defined as sharing a common point or boundary edge of the unit polygon. This is accomplished in (5–13) by establishing a sink unit among a patch, then structuring flow from a selected unit that must traverse through other selected units and end at the sink. Murray and Church [23] offer proofs of model completeness, including that contiguity among units in a selected patch is enforced, p patches will be identified and that the limit on maximum average size of a patch is maintained, among other properties.

The multi-objective extension reported in Murray et al. [21] is important for a number of reasons, but especially in demonstrating how many considerations can be simultaneously considered in the context of fuels management prioritization in and around the wildland urban interface. The analysis was for coastal Santa Barbara County consisting of 26,326 parcels (out of 41,087 in the region) within the wildland urban interface, with each unit being considered in the model application for some type of fuels mitigation. The scenario reported here involves identifying 64 patches/ project areas with the condition that each patch be a contiguous sub-area and no larger than 100 acres on average. Analysis was carried out on a personal computer (Intel i7, 3.61 GHz, 32 GB RAM) running Windows 10, with model implementation in Jupyter Notebook (Python 3.7) using Gurobi 9.5 to solve all optimization problem instances. The β_{i1} values in the first objective, (16), are a measure of wildfire hazard derived from FlamMap simulations as a function of conditional flame length and burn probability. The β_{i2} values in the second objective, (17), are a measure of social vulnerability should wildfire impact a parcel unit. The constraint method was used to deal with multiple objectives by moving (17) to the constraint set bounded to be above a stipulated threshold. Solution time for each problem instance was limited to one hour. 71 non-dominated solutions were identified by incrementally varying the social vulnerability threshold, with an average optimality gap of 7.50%. One non-dominated solution is shown in Figure 2, with the inset showing the tradeoff frontier of the 71 solutions. The map in Figure 2 corresponds to the highlighted tradeoff solution with the 64 selected patches indicated across the region.

It is clear in Figure 2 that a range of tradeoff alternatives exist in this case for wildfire mitigation, each addressing a different level of fire hazard and social vulnerability. Each of the 71 tradeoff solutions are viable to consider, representing a compromise between these two objectives. Worth noting in Figure 2 is that the efficiency frontier is non-convex. Further, the spatial pattern for this one compromise solution is rather dispersed, but also concentrated in particular areas. Of course, spatial pattern can vary substantially between tradeoff solutions, but in particular as a result of increasing or decreasing the importance of social vulnerability and/or hazard potential.

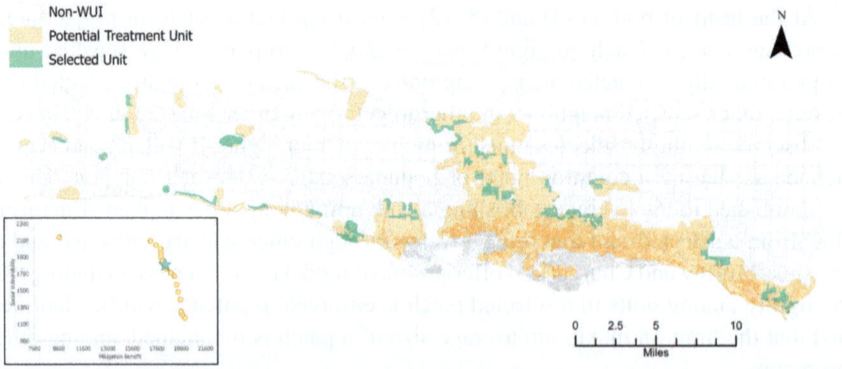

Fig. 2 Tradeoff curve and spatial configuration of one solution

4 Challenges

There are many challenges for enhancing support of planning, management, decision making and/or policy associated with mitigating risk and vulnerability of extreme event wildfire. In what follows, discussion of spatial information, wildfire behavior, model extensions and supporting model solution analytics are touched upon, but certainly others could be included as well.

4.1 Spatial Information

The ability to aid analysis and decision making processes is dependent on a reasonable understanding of needs along with sufficient data that meaningfully represents observed conditions and relationships. In the context of wildfire fuels mitigation described previously, at a minimum detailed spatial information on fuels in a parcel is necessary as are details associated with the social vulnerability of those residing in or frequenting a parcel. Viewed in one way, this type of information does exist and can be brought to bear in risk mitigation (see Figure 2). However, conditions can change rapidly, with observed and/or derived data often not keeping pace. As an example, vegetation conditions are very much impacted by weather and climate. Is the relied upon data sufficiently reflecting the actual or expected state of vegetation? Does the resolution of observed/forecast vegetation reflect the scale of analysis being undertaken? What about other fuels that contribute to wildfire risk? In terms of social vulnerability, it is a challenge to identify those that would be inordinately impacted by wildfire, having less capacity to recover and/or respond to an incident. While traditional census data is one likely source of information for identifying the vulnerable, e.g., young, old, impoverished, unemployed, racial/ethnic minority, etc. reflected in Figure 2, such details are only available at coarse spatial resolutions. This makes interpretation difficult at the decision making unit level such as the parcel. Additionally, what about the people not represented in census information, such as the unhoused? There are also mobility considerations, and whether information collected years earlier actually reflects current or anticipated conditions. Much more could be said about spatial information, but the point here is simply to highlight aspects of data quality and representativeness that are often not taken into account, or even acknowledged.

4.2 *Wildfire Behavior*

There has been much effort and interest in developing and applying wildfire simulation approaches, something that is critical for estimating risk and exposure. Much of this has been focused on forested regions. Wildfire behavior is very different in non-forest regions, however. Mediterranean climate regions, e.g., southern California, Greece, etc., are a particular concern as they continue to experience major increases in extreme wildfires. Further, the understanding of fire behavior dynamics is not particularly high, especially in and around the wildland urban interface. While this could be noted in the previous sub-section, weather and climate information are often represented in very coarse resolutions, lacking in important spatial detail needed for better prediction of fire behavior. Finally, extreme conditions that drive catastrophic wildfire incidents remain challenging to understand for many reasons, but the movement of embers and firebrands is particularly confounding as they are largely driven by local and erratic fire induced winds.

4.3 *Model Extensions*

It is clear that fuels reduction through a variety of interventions can do much to reduce extreme wildfire risk (see Calkin et al. [7]; Finney [8]). Modeling has therefore centered on identifying the most strategic patches (or project areas) for mitigation efforts based upon derived benefits. Developed models and support systems work represent a significant advance with proven results. Extensions and considerations that focus on more explicitly accounting for the interactions of collective efforts, including fuels reductions, firebreaks, water/retardant drops, etc., in a more dynamic manner represent an important area for future work. That is, spatial configuration matters, and likely is non-linear in nature based upon the interactions of various mitigation efforts.

As noted above, patches are conceived to be administrative in nature due to where a project is to be undertaken, and are often limited in size based on management and mitigation implementation. Because of this, patches may neighbor each other in practice. However, there are situations where this may not be desirable as the combination of two or more patches could represent an excessive local disturbance area. Extension to address this might include approaches along the lines of those relied upon in harvest scheduling, such as [24, 25].

4.4 Supporting Model Solution Analytics

As noted previously, much has been done to support the theory and practice of advanced analytics to help make the best decisions possible in wildfire risk mitigation. This includes the development of heuristics and exact approaches to identify good (or the best) management solutions. The identification of tradeoffs between competing objective criteria too is a critical area of essential support. Pludow et al. [19] considered multi-objective evaluation using ForSys to maximize resilience and minimize expected loss. They demonstrated that heuristic results, while relatively good when viewed in isolation, are potentially problematic due to suboptimalty combined with limited ability to identify the actual range of tradeoff solutions that exist.

Why care about multiple objectives and the ability to identify solutions that define and characterize the Pareto frontier? First, with multiple objective problems there is no single optimal solution. Rather, there exist Pareto tradeoffs that represent a compromise of one or more objectives in order to enhance another objective. Given the above model, let Z_1 equal value of (5) and Z_2 equal value of (6) for a given feasible solution. A non-dominated solution Z_1'' and Z_2'' is defined based on one of these conditions for any solution Z_1' and Z_2':

$$Z_1' \leq Z_1'' \quad \text{and} \quad Z_2' > Z_2'' \tag{14}$$

or

$$Z_1' < Z_1'' \quad \text{and} \quad Z_2' \geq Z_2'' \tag{15}$$

In practice there are a range of methods used to obtain tradeoff solutions. A number of different approaches for solving multi-objective problems are detailed in Cohon [26]. The weighting and constraint methods are relied upon in practice. The weighting method, e.g., (1), involves the use of a priori defined weights to give priority to the different objectives. Typically the weights are conditioned as follows:

$$\sum_{k \in K} w_k = 1 \tag{16}$$

Different weights can be evaluated and the model re-solved. Of course, an infinite number of weight combinations will theoretically need to be considered, unless model properties can be established for systematically altering weights. An additional complication with the weighting method is that it cannot find unsupported solutions when the decision variable are non-continuous, meaning the the Pareto frontier could only be approximated. As an alternative, the constraint method can find all Pareto solutions, supported and unsupported. Solution using the constraint method requires the additional objective(s) to be moved to constraints that are parameterized. For the above model, this could involve keeping (5) and moving (6) to a constraint. This would be accomplished as follows:

$$\sum_{i \in I} \beta_{i2} X_i \leq \delta \tag{17}$$

where δ is an a priori limit on objective (2). Then a range of values $\delta \in [\delta_{min}, \delta_{max}]$ can be considered within a process that solves the model for each unique value of δ. This was in fact the approach used to derive tradeoff solutions reported in Figure 2.

The point is that the weighting method may be limiting in practice, but also that heuristic solution approaches must be designed to identify a complete range of tradeoff solutions to more effectively support planning and decision making processes. This is precisely the take home message in Pludow et al. [19]. Work by Xu et al. [27] has proposed measures and metrics for characterizing success in tradeoff solution identification when heuristics are used, and were the basis of the assessment carried out in Pludow et al. [19].

5 Conclusions

The devastation and destruction of wildfire is unimaginable, something that the recent wildfires in Los Angeles have reminded us. Wildfire resilience and coexistence is something that will not achieved by chance. To support purposeful actions that will reduce fuels and decrease the risk of extreme wildfire events, spatial analytics and models have been developed and applied broadly. The work by [2, 9, 16, 20, 23] is noteworthy along these lines (see [Ager 22]) associated with patch contiguity and size needs.

Something that is often lost in the presentation and application of either heuristic techniques and exact methods in wildfire mitigation is that there is actually an important role for both approaches in practices. Ager [22] highlights the computational efficiency of heuristics to support interactive decision making processes, often identifying solutions in seconds. However, exact methods based on mixed integer programming are also fundamentally important as they offer a basis for comparison and quality assessment in results obtained from heuristic techniques. Since one never knows how good a heuristic solution is, establishing quality bounds along these lines is fundamental for good decision making. Beyond this, the multiple objective nature of wildfire mitigation means that care must be taken to identify and consider the full range of tradeoffs that exist. This will not happen by chance, so care must be taken to ensure they are found and considered.

Extensions and continued research are essential. The need will always be high for better, faster heuristics in dynamic decision making environments. Exact methods with improved solution properties capable of faster solution that accommodate larger problem instances also will continue to be critical. As discussed above, the emergence of new problem constructs, heuristic and exact, that reflect the dynamic nature wildfire and mitigation, accounting for interactions of whatever form, are no doubt on the horizon. This will require the ultimate integration of wildfire and spatial optimization.

Acknowledgements The generous support of Yardi Systems through the Wildfire Resilience Initiative at University of California, Santa Barbara is gratefully acknowledged.

References

1. Balch, J.K., Iglesias, V., Mahood, A.L., Cook, M.C., Amaral, C., DeCastro, A., Leyk, S., McIntosh, T.L., Nagy, R.C., St. Denis, L., et al.: The fastest-growing and most destructive fires in the us (2001 to 2020). Science **386**(6720), 425–431 (2024)
2. Ager, A.A., Evers, C.R., Day, M.A., Alcasena, F.J., Houtman, R.: Planning for future fire: scenario analysis of an accelerated fuel reduction plan for the western United States. Landsc. Urban Plan. **215**, 104212 (2021)
3. Moritz, M.A., Batllori, E., Bradstock, R.A., Gill, A.M., Handmer, J., Hessburg, P.F., Leonard, J., McCaffrey, S., Odion, D.C., Schoennagel, T., et al.: Learning to coexist with wildfire. Nature **515**(7525), 58–66 (2014)
4. Radeloff, V.C., Mockrin, M.H., Helmers, D., Carlson, A., Hawbaker, T.J., Martinuzzi, S., Schug, F., Alexandre, P.M., Kramer, H.A., Pidgeon, A.M.: Rising wildfire risk to houses in the United States, especially in grasslands and shrublands. Science **382**(6671), 702–707 (2023)
5. Syphard, A.D., Velazco, S.J.E., Rose, M.B., Franklin, J., Regan, H.M.: The importance of geography in forecasting future fire patterns under climate change. Proc. Natl. Acad. Sci. **121**(32), e2310076121 (2024)
6. Parkins, K., Cawson, J., Pickering, B., Penman, T.: Mitigation strategies for wildfires. In: Meacham, B.J., McNamee, M. (eds.) Handbook of Fire and the Environment: Impacts and Mitigation, pp. 395–420. Springer International Publishing, Cham (2023)
7. Calkin, D.E., Cohen, J.D., Finney, M.A., Thompson, M.P.: How risk management can prevent future wildfire disasters in the wildland-urban interface. Proc. Natl. Acad. Sci. **111**(2), 746–751 (2014)
8. Finney, M.A.: Design of regular landscape fuel treatment patterns for modifying fire growth and behavior. Forest Sci. **47**(2), 219–228 (2001)
9. Ager, A.A., Vaillant, N.M., McMahan, A.: Restoration of fire in managed forests: a model to prioritize landscapes and analyze tradeoffs. Ecosphere **4**(2), 1–19 (2013)
10. Chung, W., Jones, G., Krueger, K., Bramel, J., Contreras, M.: Optimising fuel treatments over time and space. Int. J. Wildland Fire **22**(8), 1118–1133 (2013)
11. Kim, Y.-H., Bettinger, P., Finney, M.: Spatial optimization of the pattern of fuel management activities and subsequent effects on simulated wildfires. Eur. J. Oper. Res. **197**(1), 253–265 (2009)
12. Pais, C., Carrasco, J., Moudio, P.E., Shen, Z.-J.M.: Downstream protection value: Detecting critical zones for effective fuel-treatment under wildfire risk. Comput. Oper. Res. **131**, 105252 (2021)
13. Schroder, S.A.K., Toth, S.F., Deal, R.L., Ettl, G.J.: Multi-objective optimization to evaluate tradeoffs among forest ecosystem services following fire hazard reduction in the deschutes national forest, USA. Ecosyst. Serv. **22**, 328–347 (2016)

14. Wei, Y., Ridcout, D., Kirsch, A.: An optimization model for locating fuel treatments across a landscape to reduce expected fire losses. Can. J. For. Res. **38**(4), 868–877 (2008)
15. Qi, J., Zhuang, J.: A review of optimization and decision models of prescribed burning for wildfire management. Risk Analysis (2024)
16. Ager, A.A., Day, M.A., Vogler, K.: Production possibility frontiers and socioecological tradeoffs for restoration of fire adapted forests. J. Environ. Manage. **176**, 157–168 (2016)
17. Borges, J.G., Marques, S., Garcia-Gonzalo, J., Rahman, A.U., Bushenkov, V., Sottomayor, M., Carvalho, P.O., Nordström, E.-M.: A multiple criteria approach for negotiating ecosystem services supply targets and forest owners' programs. Forest Sci. **63**(1), 49–61 (2017)
18. Ferreira, L., Nascimento Baptista, A., Constantino, M., Marques, S., Martins, I., Borges, J.G.: Integrating wildfire resistance and environmental concerns into a sustainable forest ecosystem management approach. Front. Forests Global Change **6**, 1177698 (2023)
19. Pludow, B.A., Murray, A.T., Echeverri, V., Church, R.L.: Evaluation of forest treatment planning considering multiple objectives. J. Environ. Manage. **346**, 118997 (2023)
20. Murray, A.T., Church, R.L., Pludow, B.A., Stine, P.: Advancing contiguous environmental land allocation analysis, planning and modeling. J. Land Use Sci. **17**(1), 572–590 (2022)
21. Murray, A.T., Baik, J., Figueroa, V.E., Rini, D., Moritz, M.A., Roberts, D.A., Sweeney, S.H., Carvalho, L.M., Jones, C.: Developing effective wildfire risk mitigation plans for the wildland urban interface. Int. J. Appl. Earth Obs. Geoinf. **124**, 103531 (2023)
22. Ager, A.: Improving the evaluation of spatial optimization models for prioritizing landscape restoration and wildfire risk reduction investments. J. Environ. Manage. **360**, 121001 (2024)
23. Murray, A.T., Church, R.L.: Spatial optimization of multiple area land acquisition. Comput. Oper. Res. **153**, 106160 (2023)
24. Goycoolea, M., Murray, A.T., Barahona, F., Epstein, R., Weintraub, A.: Harvest scheduling subject to maximum area restrictions: exploring exact approaches. Oper. Res. **53**(3), 490–500 (2005)
24. Martins, I., Alvelos, F., Cerveira, A., Kašpar, J., Marušák, R.: Solving a harvest scheduling optimization problem with constraints on clearcut area and clearcut proximity. Int. Trans. Oper. Res. **30**(6), 3930–3948 (2023)
26. Cohon, J.L.: Multiobjective programming and planning (New York: Academic Press). (1978)
27. Xu, J., Murray, A.T., Church, R.L., Wei, R.: Service allocation equity in location coverage analytics. Eur. J. Oper. Res. **305**(1), 21–37 (2023)

Covering with Network Design for Wildfire Promptness

Elsa Silva, Filipe Alvelos, and Marco Marto

Abstract We consider the problem of selecting bases for firefighting activities (e.g., vigilance, water refill, initial attack) and links between them in the context of wildfire promptness. Bases can be facilities, such as watchtowers and water tanks, or positions from where an initial attack is conducted. It is assumed that it is advantageous to connect bases in such a way that resources (e.g. ground crews) can quickly move between them. The general problem is modelled in a general way as integration of a set covering problem (for selecting the location of the bases) and a travelling salesman problem where the cities are the selected locations and the arcs the links that connect them. We propose a mixed integer programming model where objectives are addressed by lexicographic optimization. The first objective is related to cover potential ignition points with a high estimate of their initial spread rate of the fire at the detection time. Computational experiments are discussed for a scenario, of an actual landscape, with parameters estimated from a fire behaviour model that takes into account slope, fuels, and wind.

Keywords Travelling salesman problem · Mixed Integer Programming · Initial fire spread simulation

E. Silva (✉) · F. Alvelos
Department of Production and Systems/ALGORITMI Research Centre/LASI, School of Engineering, University of Minho, Braga, Portugal
e-mail: elsa@dps.uminho.pt

F. Alvelos
e-mail: falvelos@dps.uminho.pt

E. Silva · M. Marto
ALGORITMI Research Center/LASI, School of Engineering, University of Minho, Braga, Portugal

F. Alvelos et al. (eds.), *Advances in Optimization and Wildfire*, Lecture Notes in Operations Research, https://doi.org/10.1007/978-3-032-03108-2_13

1 Introduction

Wildfires pose an increasing threat to ecosystems, human communities, and infrastructure worldwide [1]. The increasing frequency and intensity of wildfires, exacerbated by climate change, highlights the urgent need for effective strategies to mitigate their impacts. Proficient wildfire management requires not only the ability to predict fire behaviour, but also the strategic deployment of resources to minimize damage. One of the most critical challenges in this context is determining optimal locations for resources, such as watchtowers, fire detection systems, or firefighting crews and ensuring efficient movement of these resources across the landscape. This challenge becomes particularly acute in regions where response times are constrained by the spatial distribution of resources and the availability of travel routes.

Wildfire management strategies can be classified into five primary categories [2]: (1) Suppression: controlling wildfires after detection with human and mechanical resources; (2) Pre-suppression: preparing and deploying resources before a fire occurs; (3) Structure protection: reducing structural damage using fire resistant materials and vegetation; (4) Fuel management: modifying vegetation to lower fire risk through prescribed burns and ecological fire use; (5) Ignition prevention: reducing wildfire ignitions through education, legal measures and regulations.

Operational research and management science have been extensively applied to improve fire management systems, as reviewed by Martell [3]. This review highlighted the development of decision support systems for fire managers, while also identifying significant gaps in the literature, particularly in optimizing the allocation of prevention resources to reduce human-caused fires.

In the domain of wildfire detection, several methods have been explored, including human observation, satellite systems, airborne systems, camera networks, and wireless sensor networks [4]. These methods are evaluated based on cost, efficiency, false alarm rates, coverage, spatial accuracy, detection delay, and the provision of fire-related information.

In Dunn et al. [5] a framework is proposed to improve decision-making in large fire management, focusing on incidents lasting more than three days and covering more than 120 hectares. It aims to integrate decision making levels in land and fire management organizations, facilitating adaptive and risk-informed strategies based on evolving conditions and resource availability. The framework stresses empirical evaluation and collaboration between decision makers and researchers to fill research gaps, improve efficiency and safety, and optimize resource allocation amid growing wildfire challenges.

Recent studies have focused on integrating advanced optimization techniques into wildfire management. For example, Dunn et al. [5] proposed a decision-making framework for large fire incidents, emphasizing adaptive, risk-informed strategies based on evolving conditions and resource availability. Similarly, Fuente et al. [6] proposed an optimization-based approach to design an integrated forest fire monitoring system to determine where to locate watchtowers and balloons, and how to route surveillance drones to cover the terrain. A mixed integer programming (MIP) model is presented that simultaneously handles location and routing decisions while minimizing the costs of building watchtowers, installing balloons, and purchasing and routing surveillance drones. Since the model is difficult to solve, a matheuristic is proposed. Computational experiments are performed with synthetic instances and a real case study in Chile.

In the broader context of location and routing problems, Drexl and Schneider [7] and Arslan [8] provided comprehensive overviews of progress in these areas. These works emphasize the integration of facility location and vehicle routing to optimize logistics planning, highlighting the importance of coverage range, vehicle capacity, and cost minimization in achieving optimal solutions.

More specifically, several studies have proposed integrated models combining coverage and routing decisions. Current et al. [9] introduced the Maximum Covering/Shortest Path Problem, extending classical coverage models to incorporate network design and routing considerations. Current and Schilling [10] developed the Median Tour and Maximal Covering Tour Problems, formulating bicriterion models to balance demand coverage and route length. In the context of strategic planning for transit networks, Matisziw et al. [11] proposed a model for extending existing routes to maximize coverage while minimizing deviation from current infrastructure. Florios and Mavrotas [12] addressed exact Pareto front generation for Multi-Objective Traveling Salesman and Set Covering Problems, relevant for applications with competing objectives. More recently, Moradi et al. [13] provided a comprehensive review and classification of Set Covering Routing Problems (SCRPs), offering a unified modelling framework that integrates coverage and routing through both direct visits and proximity-based assignment. These studies provide important foundations and justify the relevance of integrated models such as the one proposed in this work, which combines spatial coverage, fire propagation risk, and circuit efficiency in a multi-objective framework tailored to wildfire response planning.

Building on this body of work, we present a MIP model that integrates the covering problem through the selection of nodes and the network design, which determines the arcs interconnecting the nodes. This integrated approach addresses critical gaps in wildfire management by optimizing resource allocation and movement within a given landscape.

Section 2 defines the problem addressed in this work. Section 3 then presents a MIP formulation to solve the problem, providing a detailed description of the parameters, decision variables, and constraints incorporated in the model. The objectives considered and the proposed solution approach are outlined in Sect. 4. Section 5 discusses the computational experiments conducted to validate the solution approach. Finally, Sect. 6 summarizes the main conclusions and outlines directions for future research.

2 Problem Definition

The problem addressed in this paper involves the design of a network where the nodes represent facilities/positions, generally termed *bases*, and the arcs represent links between them. The location of a base is assumed to reduce the impact of an ignition in the surrounding area.

The landscape is assumed to have been premapped and modelled with detailed spatial, topographic and fuel-level data. Using this information, it is possible to predict expected fire-related measures such as the fire rate of spread (ROS) in different directions, after an ignition occurs. This prediction enables informed decision making based on the minimum travel-time principle, which assumes that the fire spreads along the shortest path in terms of transmission times. Details on how landscape, fire, and resources are characterized can be found in Alvelos [14, 15].

The landscape is modelled as a set of points representing potential ignitions and a network for the positions and movements of the resources. The nodes of this network represent potential locations for installing bases for resources such as surveillance towers, fire detection stations, or positions of firefighting crews to attack fire. The arcs represent links, such as roads or trails, that allow the movement of resources between bases. The problem consists in selecting a given number of positions for bases of different types to cover the potential ignitions and a set of roads linking the selected bases.

For example, the resources may be firefighting crews and the bases suitable positions to conduct initial attacks, being appropriate to have fast links between them. Another example is detection; bases are suitable locations for installing watchtowers to detect ignitions as early as possible. If the human resources available for the detection are not enough to keep all the watchtowers active, it is advisable that the existent ones can move quickly between bases. A last example is surveillance; the resources are patrols, and the bases are positions that deserve particular attention. In order to increase the surveillance at those points, the time of traveling in the links between them should be as low as possible.

Fire, which can be seen as the demand in the well-known facility location problem, is modelled by associating a potential ignition to each point in a grid as illustrated in Fig. 1, where the (road) network associated with the resources is also displayed.

The goal of the problem is twofold:

- **Covering Problem**: Determine the optimal positions of the bases to maximize the coverage of the potential ignition nodes, minimizing the impact of fire spread (because of early detection or early attack). Each base has a predefined coverage radius, and, ideally, every potential ignition node is covered by at least one base.
- **Network Design Problem**: Designing a network in such a way that it allows efficient travel between the selected bases, minimizing the total travel distance for resources to move between bases, while forming a circuit, assuring two disjoint paths between any pair of bases.

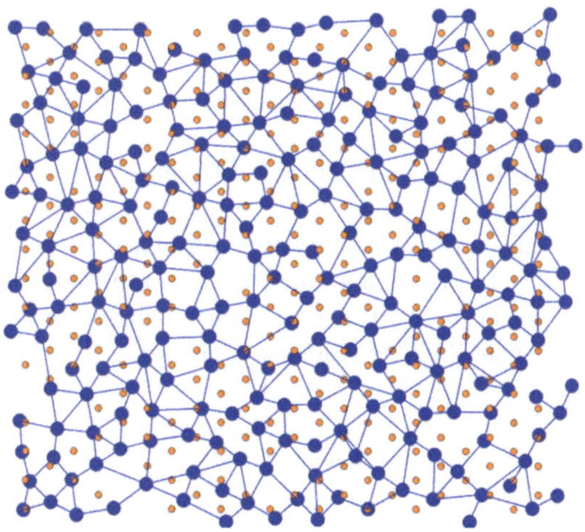

Fig. 1 Ignition points and resource network representation. The blue nodes represent locations where resources can be assigned, and the arcs represent the connections between these nodes, such as roads. The orange nodes represent potential ignition points. Missing ignition points in the grid mean that the corresponding cell is unburnable

To estimate the relative importance of covering each ignition point, an estimate of the accumulated initial ROS is used, that is, the sum of the ROS in each active arc (obtained in Alvelos [14]), as illustrated in Fig. 2.

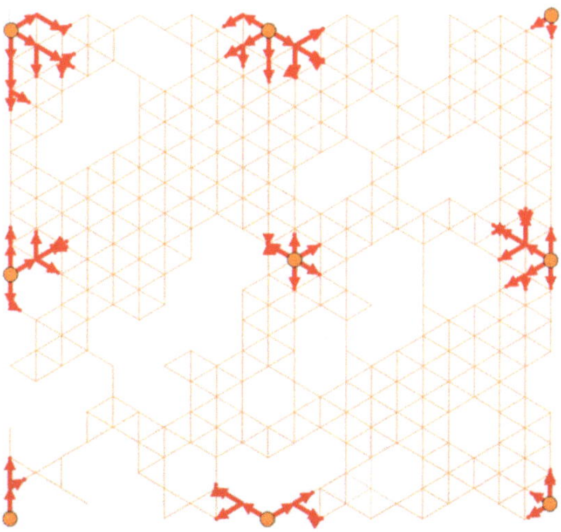

Fig. 2 Examples on the state of fire when detected, after 30 minutes of the ignition, for 9 ignitions with a upslope wind at 30 km/h

3 Mixed Integer Programming Model

This section presents the parameters, decision variables and constraints of a MIP that integrates the covering problem (for selecting the positions of the bases) and the network design problem (for selecting the arcs and connecting the selected nodes in a circuit).

The objective functions are introduced in the following section.

The formulation for the covering problem is based on a *p*-median location model, while the network design is inspired by the model proposed for the Travelling Salesman Problem in Miller et al. [16].

Parameters :

N_d : set of potential ignition nodes

ROS_j : accumulated fire rate of spread of ignition point $j, j \in N_d$

N_b : set of potential base nodes

B - set of types of bases

p^b : number of bases of type b that may be selected, $b \in B$.

$$R_{ij}^b = \begin{cases} 1, & \text{if ignition } j \text{ is in the radius of base } i \text{ of type } b, j \in N_d, n \in N_b, b \in B \\ 0, & \text{otherwise.} \end{cases}$$

A_b : set of arcs directly linking all pairs of nodes of the network

t_{ij} : duration of the quickest path from base i to base j, $ij \in A_b$

c^b : cost of a base of type b, $b \in B$

Parameters t_{ij} are obtained by solving an all-pairs shortest paths problem (in fact, quickest paths as the weights of the arc are times) in the road network (with different speeds for different types of roads) and therefore have the duration between each pair of nodes of the base nodes.

Decision Variables :

$$v_j = \begin{cases} 1, & \text{if potential ignition node } j \text{ is covered} \quad j \in N_d \\ 0, & \text{otherwise.} \end{cases}$$

$$w_{ij}^b = \begin{cases} 1, & \text{if node } i \text{ of base type } b \text{ is selected to cover potential ignition node } j, \\ & \quad i \in N_b, j \in N_b, b \in B \\ 0, & \text{otherwise} \end{cases}$$

$$y_i^b = \begin{cases} 1, & \text{if a base of type } b \text{ is located at node } i, \quad i \in N_b, b \in B \\ 0, & \text{otherwise.} \end{cases}$$

$$z_i = \begin{cases} 1, & \text{a base of any type is located at node } i, \quad i \in N_b \\ 0, & \text{otherwise.} \end{cases}$$

$x_{ij} \quad = \begin{cases} \text{if arc } ij \text{ is selected} & ij \in A_b \\ 0, \text{ otherwise.} \end{cases}$

u_i : position of node i in the sequence of visited bases $i \in N_b \backslash \{0\}$

The constraints are the following.

$$v_j \leq \sum_{b \in B} \sum_{i \in N_b} R^b_{ij} w^b_{ij} \qquad j \in N_d \qquad (1)$$

$$R^b_{ij} w^b_{ij} \leq y^b_i \qquad i \in N_b, j \in N_d, b \in B \qquad (2)$$

$$\sum_{i \in N_b} y^b_i \leq p^b \qquad b \in B \qquad (3)$$

$$y^b_i \leq z_i, \qquad i \in N_b, b \in B \qquad (4)$$

$$\sum_{j \in \delta^+(i)} x_{ij} = z_i \qquad i \in N_b \qquad (5)$$

$$\sum_{j \in \delta^-(i)} x_{ji} = z_i \qquad i \in N_b \qquad (6)$$

$$u_j \geq |N_b|\big(x_{ij} - 1\big) + u_i + 1 \qquad ij \in A_b, i \neq 0, j \neq 0 \qquad (7)$$

$$0 \leq u_i \leq |N_b| \qquad i \in N_b \backslash \{0\} \qquad (8)$$

$$v_j, w^b_{ij}, y^r_i \in \{0, 1\}, \qquad j \in N_d, i \in N_b, b \in B, \qquad (9)$$

$$x_{ij} \in \{0, 1\}, \qquad ij \in A_b \qquad (10)$$

$$u_i \geq 0 \text{ and integer} \qquad i \in N_b \backslash \{0\} \qquad (11)$$

Constraints (1)–(3) define the covering problem. Constraints (1) ensure that if the the potential ignition node j is covered, must have at least a base of a type with a radius greater than their distance. Constraints (2) guarantee that if a node is used to cover an ignition it must have a base. The limitation on the number of bases of each type is enforced by constraints (3). Constraints (4) allow the identification of the nodes with at least one base, the ones that must be linked.

Constraints (5)–(9) address the network design problem. Constraints (5) and (6) ensure that the selected nodes for the bases that are part of the circuit. The sequence in which the base nodes are visited, as well as the elimination of subtours, is enforced by constraint (7). Constraints (8) limit the index of the nodes to be visited. Finally, the domain of each decision variable is specified in constraints (9)–(11).

4 Objectives and Solution Approach

Different objective functions are considered. Two of them are related with covering and take into account the accumulated ROS at the uncovered potential ignition nodes. Objective function (12) minimizes the total accumulated ROS across all uncovered potential ignition nodes, while objective function (13) minimizes the maximum ROS among the uncovered potential ignition nodes, requiring an additional set of constraints (14).

$$\text{Minimize z} = \sum_{j \in N_d} (1 - v_j) \cdot \text{ROS}_j \tag{12}$$

$$\text{Minimize z} = \Psi \tag{13}$$

Subject to:

$$(1 - v_j) \cdot \text{ROS}_j \leq \Psi \qquad j \in N_d. \tag{14}$$

Objective (15) focuses on maximizing the number of potential ignitions covered.

$$\text{Maximize z} = \sum_{j \in N_d} v_j \tag{15}$$

Objective (16) minimizes the cost associated with the types of the selected bases.

$$\text{Minimize z} = \sum_{b \in B} \sum_{i \in N_b} c^r \cdot y_i^b \tag{16}$$

Regarding the network design problem, a single objective function is considered, which aims to minimize the duration of the circuit connecting all bases, as defined in Ehrgott (17).

$$\text{Minimize z} = \sum_{ij \in A_b} t_{ij} \cdot x_{ij} \tag{17}$$

We used lexicographic optimization. This method addresses objectives sequentially, establishing a hierarchy of objective functions prioritized according to their importance, without the need for weighting factors [17]. Each optimization step progressively narrows the solution space, using subsequent objectives to distinguish between alternative optimal solutions.

The objectives are optimized sequentially, according to a predefined priority. Let X be the feasible set. The optimization proceeds as follows:

- $X_1 = \arg\min_{x \in X} f_1(x)$
- $X_2 = \arg\min_{x \in X_1} f_2(x)$
- $\ldots$
- $X_k = \arg\min_{x \in X_{k-1}} f_k(x)$

The final solution is any $x^* \in X_k$. This ensures that each objective is optimized without compromising the optimality of higher-priority objectives.

We considered two alternative first objectives, (12) or (13), followed by the maximization of the number of covered potential ignitions (15), the minimization of the costs associated with the bases (16), and the minimization of the duration of the bases network (17).

Objectives (12) and (13) integrate the information on the ROS of the fire, assigning greater importance to the ignition points where the spread of the fire is expected to be faster. This risk-aware perspective is particularly relevant in scenarios with limited suppression resources, where strategic prioritisation is necessary. In contrast, objective (15) considers only the number of ignition points covered, without accounting for fire behaviour, thus providing a purely spatial, data-independent baseline. Including these formulations allows us to compare alternative strategies, some informed by fire dynamics, others by spatial distribution alone, and assess their implications in different practical settings.

In a few tests, the MIP model described in the previous section took too much time to reach optimality, in particular in closing the gap of the fourth objective (17). While the MIP deserves more work, in particular exploring incumbent solutions, we decided to conceive a two-step heuristic to validate the approach and obtain a set of preliminary results.

In the first step of the two-step heuristic, the covering objectives, (12) or (13), followed by (15) and (16) are lexicographically optimized. This optimization results in locations for the bases. In the second step, a TSP is solved with the cities corresponding to the selected locations for bases in the first step.

5 Computational Experiments

In this study, we consider the Baião region in Portugal, with an extent of $14 \times 18\text{km}^2$, as the landscape under study, using pyO3F—A python framework for wildfire-related optimization [14, 15].

This framework uses information on the terrain (i.e. elevations), fuels and wind to characterize a network where nodes correspond to (hexagonal) cells and arcs correspond to potential fire transmissions between adjacent cells. Unburnable cells (e.g. where water is dominant) are not translated in nodes (as can be seen in Fig. 2). Different resolutions, i.e. distances between adjacent nodes, can be used. Each arc has an estimate of the fire transmission time as an attribute. Arcs are directed as the fire transmission time depends on slope. Once, at least, one ignition occurs an estimate of the ROS between any two nodes of the landscape can be derived through quickest paths, according to the fire minimum travel time principle. The potential locations and movements for resources are defined in a different network that depend on the type of resource. For example, ground crews are associated with a network representing roads and paths (as can be seen in Fig. 1). More details can be found in the already mentioned papers [14, 15].

Fire is represented on a hexagonal grid with a resolution of 1000 meters resulting in 310 potential ignition nodes. The resource network has 326 nodes (we assume all of them are potential locations) and 1276 arcs. For the resolution of 500 meters 1078 nodes and 3488 arcs were considered. Three types of bases are considered, each with a distinct coverage radius. The bases have the following coverage radius: 4000 meters, 3000 meters, and 2000 meters, corresponding to unit costs of 4, 2, and 1, respectively. For each potential ignition node $j \in N_d$, the accumulated fire ROS is obtained from the predicted fire status at the detection time. In Fig. 3 the potential ignitions are presented with colors representing the accumulated ROS.

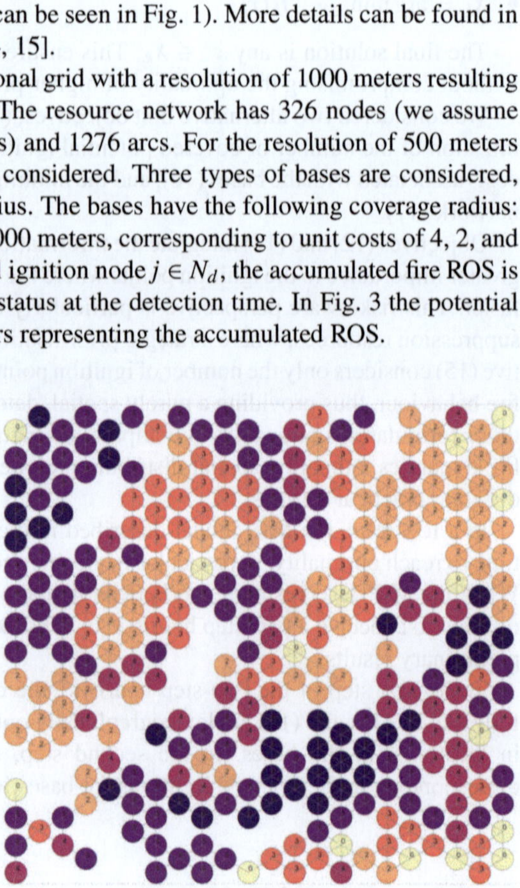

Fig. 3 Ignition points. Darker colors correspond to higher ROS values

5.1 *Influence of Considering* Min_Sum *or* Min_Max *as the First Objective in Step 1*

The computational tests were conducted on a machine with a processor Intel Core i7, 1700 Mhz, 10 cores, 12 logical processors and 32 GB of physical memory.

The first scenario evaluated in the computational experiments examined the effect of including the objectives (12) or (13) in what concerns the ROS of fire at the uncovered potential ignition nodes, as part of the first objective in step 1 of the two-step approach.

In Fig. 4, the results obtained by considering objective (12) as the first objective in step 1 of the solution approach are shown on the left side, while the results obtained with objective (13) are displayed on the right side. In these experiments, it was considered that 2 bases of each type could be established.

The red circles represent bases with a coverage radius of 4000 meters, the yellow circles represent those with a radius of 3000 meters, and the grey circles correspond to bases with a radius of 2000 meters. The blue line illustrates the circuit connecting all nodes where bases have been established.

The total number of available bases was insufficient to cover all potential ignition nodes. When prioritizing the *Min_Sum* objective (12), the solution resulted in a total fire ROS of 364 km/h for the uncovered potential ignition nodes. Additionally, 59% of all potential ignition nodes were covered by at least one base. The total cost was 14 ($4 \times 2 + 2 \times 2 + 1 \times 2$), and the total duration of the circuit connecting all base nodes was 89 minutes.

In contrast, when the priority was given to the *Min_Max* objective (13) in step 1 of the two-step approach, the maximum fire ROS for the uncovered potential ignition nodes was reduced to 5.2 km/h. However, only 56% of potential ignition nodes were covered, reflecting a slightly lower coverage. The total cost remained 14, as all

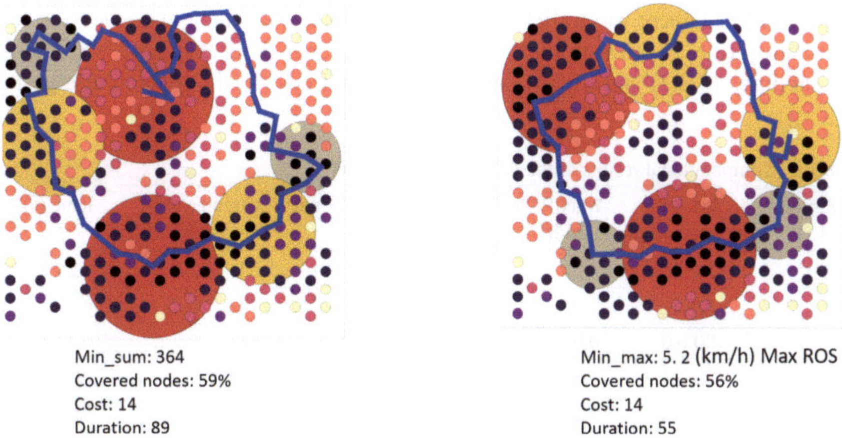

Fig. 4 Analysis of Min_Sum and Min_Max objectives in the step 1

available bases were utilized. Notably, the total duration of the circuit connecting all the base nodes was significantly reduced from 89 to 55 minutes.

For both first objectives, the computational time was less than two seconds.

Given the trade-offs observed between coverage, circuit duration, and fire propagation risk, we opted to proceed with the *Min_Max* objective in the remaining analyses. This choice reflects a prioritization of operational efficiency and risk mitigation, which are particularly relevant in scenarios with limited resources and high variability in fire behaviour. Although both formulations provide valuable perspectives, the *Min_Max* objective was deemed more suitable for the context of the Baião region and the structure of the problem addressed.

5.2 *Impact of Different Resources Availability and Resolution Grid*

In Table 1, the results of the computational experiments are presented, varying the availability of bases and the resolution of the grid. The first two columns indicate the characteristics of the problem being solved, while the remaining columns report the results obtained.

The first column shows the resolution of the fire grid considered for the Baião region, and the second column presents the availability of each base type. For instance, the entry (2, 5, 5) indicates that 2 resources with a coverage radius of 4000 meters, 5 resources with a coverage radius of 3000 meters, and 5 resources with a coverage radius of 2000 meters are considered.

Analyzing the results, it can be seen that as the number of resources increases, both the percentage of covered potential ignition nodes and the cost also increase. For the 1000-meter resolution grid, the solver demonstrated excellent performance, obtaining the optimal solution in less than 2 s. When the 500-meter resolution grid was considered, there was a slight increase in computational time, with the resource availability (2,5,5) showing the highest computational time. Instances with bases availabilities (2,2,2) and (2,5,5) had the same total cost for both resolution grids.

Table 1 Computational results of the different experiments.

Resolution (m)	Available Resources	Max ROS (km/h)	Ignitions Covered (%)	Cost	Circuit Duration (min)	Time (s)
1000	(2,2,2)	5.2	56	14	55	< 2
	(2,5,5)	3.3	86	23	168	< 2
	(20,0,0)	0.0	100	36	81	< 2
500	(2,2,2)	5.8	55	14	103	13
	(2,5,5)	5.3	81	23	195	30
	(20,0,0)	0.0	100	40	171	4

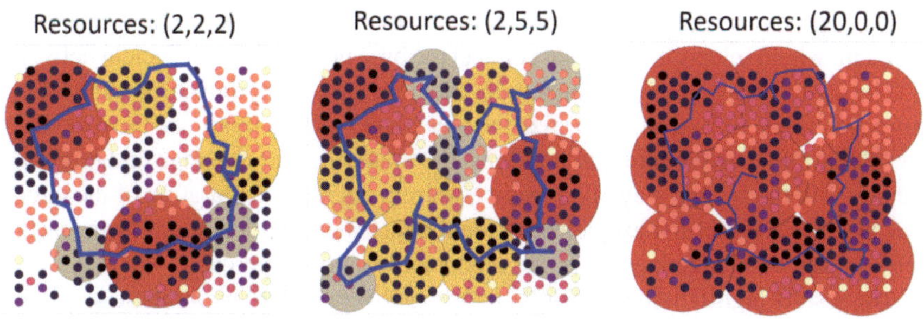

Fig. 5 Solutions for the 1000-meter resolution grid

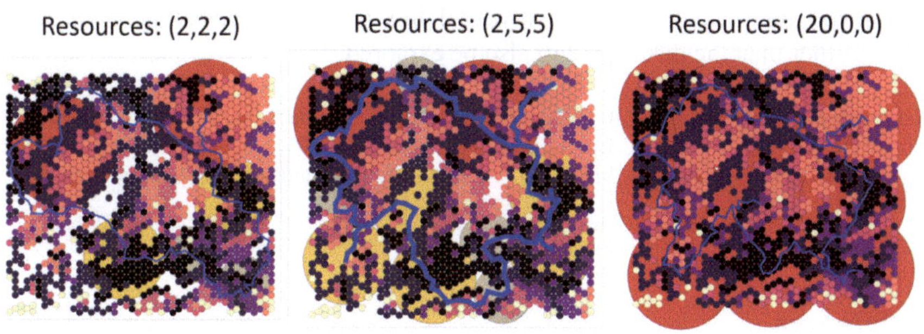

Fig. 6 Solutions for the 500-meter resolution grid

However, for the instance with availability (20,0,0) and the 500-meter resolution grid, an additional resource was needed compared to the 1000-meter resolution grid.

In Fig. 5, the solutions for different base availabilities are presented for the 1000-meter resolution grid. For bases availabilities (2,2,2) and (2,5,5), all available bases were established, while in the case with availability (20,0,0), only nine bases were sufficient to cover all the potential ignition nodes. The cases with a higher number of available bases can be used to determine the minimum number of bases required to cover all potential ignition nodes.

The solutions for different base availabilities with a 500-meter resolution grid are presented in Fig. 6. The number of potential ignition nodes is higher in this resolution grid, and the duration of the circuits also increases.

6 Conclusions

This paper addressed the problem of designing a network for installing facilities or selection positions for wildfire resources with key tasks such as surveillance, fire detection, and fire initial attack. A MIP model was proposed to integrate both the covering problem, which determines optimal base positions, and the network design problem, which selects the arcs that connect the base positions in an efficient network.

Through a two-step solution approach, a lexicographical optimization method was employed in the first step to prioritize the objectives, such as minimize the maximum fire ROS of uncovered potential ignitions, the maximize the coverage and minimize the costs. In the second step, the locations were fixed, and the duration of the circuit was minimized. Preliminary computational experiments in a real-world landscape demonstrated the effectiveness of the model and solution approach.

The results highlight the trade-offs between coverage, circuit duration, and bases costs and provide valuable information on the planning and design of wildfire resource networks.

Future work will include exhaustive computational tests. Different topologies for the network design and alternative ways of defining the coverage taking into account factors other than the distance, will also be explored.

Acknowledgements This work has been supported by FCT - Fundação para a Ciência e Tecnologia within the R&D Unit Project Scope UID/00319/Centro ALGORITMI (ALGORITMI/UM) and within project PCIF/GRF/0141/2019 "O3F - An Optimization Framework to Reduce Forest Fire".

References

1. Heyns, A.M., Plessis, W., Curtin, K.M., Kosch, M., Hough, G.: Decision support for the selection of optimal tower site locations for early-warning wildfire detection systems in south africa. Int. Trans. Oper. Res. **28**(5), 2299–2333 (2021)
2. Heineke, J.M., Weissenberger, S.: An analysis of wildfire prevention. IEEE Trans. Syst. Man Cybern. **6**, 513–522 (1974)
3. Martell, D.L.: A review of recent forest and wildland fire management decision support systems research. Curr. For. Rep. **1**, 128–137 (2015)
4. Li, Y., Zhang, T., Ding, Y., Wadhwani, R., Huang, X.: Review and perspectives of digital twin systems for wildland fire management. J. For. Res. **36**(1), 1–24 (2025)
5. Dunn, C.J., Thompson, M.P., Calkin, D.E.: A framework for developing safe and effective large-fire response in a new fire management paradigm. For. Ecol. Manage. **404**, 184–196 (2017)
6. Fuente, R., Aguayo, M.M., Contreras-Bolton, C.: An optimization-based approach for an integrated forest fire monitoring system with multiple technologies and surveillance drones. Eur. J. Oper. Res. **313**(2), 435–451 (2024)
7. Drexl, M., Schneider, M.: A survey of variants and extensions of the location-routing problem. Eur. J. Oper. Res. **241**(2), 283–308 (2015)
8. Arslan, O.: The location-or-routing problem. Transportation Research Part B: Methodological **147**, 1–21 (2021)
9. Current, J.R., Velle, C.R., Cohon, J.L.: The maximum covering/shortest path problem: A multiobjective network design and routing formulation. Eur. J. Oper. Res. **21**(2), 189–199 (1985)
10. Current, J.R., Schilling, D.A.: The median tour and maximal covering tour problems: Formulations and heuristics. Transp. Res. Part B: Methodological **73**(1), 114–126 (1994)
11. Matisziw, T.C., Murray, A.T., Kim, C.: Strategic route extension in transit networks. Eur. J. Oper. Res. **171**(2), 661–673 (2006)

12. Florios, K., Mavrotas, G.: Generation of the exact pareto set in multi-objective traveling salesman and set covering problems. Appl. Math. Comput. **237**, 1–19 (2014)
13. Moradi, N., Mafakheri, F., Wang, C.: Set covering routing problems: A review and classification scheme. Comput. Ind. Eng., 110730 (2024)
14. Alvelos, F.: pyo3f - a python framework for wildfire-related optimization, part i: design and fundamentals. In: Alvelos, F., Martins, I., Rocha, A. (eds.) Proceedings of the First Optimization and Wildfire Conference, Springer (2025)
15. Alvelos, F.: pyo3f - a python framework for wildfire-related optimization, part ii: Usage. In: Alvelos, F., Martins, I., Rocha, A. (eds.) Proceedings of the First Optimization and Wildfire Conference, Springer (2025)
16. Miller, C.E., Tucker, A.W., Zemlin, R.A.: Integer programming formulation of traveling salesman problems. J. ACM (JACM) **7**(4), 326–329 (1960)
17. Ehrgott, M.: Multicriteria Optimization, 2nd edn. Springer (2005)

An Integer Programming Model for Sensor Placement in LoRaWAN Networks

Javier Marenco, Jessica Singer, and Paula Zabala

Abstract A promising approach for the early detection of fires in forests consists on placing ground-based sensors that communicate through a *LoRaWAN network* with certain *gateways*. LoRaWAN networks have gained significant interest recently, especially with the rise of solutions based on *Internet of Things*. In this work we address the problem of sensor placement in LoRaWAN networks using inte ger programming techniques, taking into account the particular characteristics of these networks. We consider a set of *points of interest*, a set of potential locations for sensors, and a set of gateways. The problem consists in determining (a) a subset of locations where sensors are to be installed, (b) a gateway and a spreading factor associated with each sensor, and (c) a channel associated with each gateway, in such a way that the number of sensors is subject to a global limit (due to budget constraints) and a per-gateway limit (due to hardware constraints), and minimizing a combination of (i) the number of uncovered points of interest and (ii) the network interference. We provide an integer programming formulation for this problem and we explore the capabilities of a state-of-the-art solver over this formulation applied to the "Pereyra Iraola" park in Buenos Aires, Argentina. We also present theoretical results on this formulation.

Keywords Fire detection · lorawan · Integer programming

J. Marenco
Escuela de Negocios, Universidad Torcuato Di Tella, Autónoma de Buenos Aires, Argentina
e-mail: javier.marenco@utdt.edu

J. Singer (✉) · P. Zabala
Universidad de Buenos Aires, Autónoma de Buenos Aires, Argentina
e-mail: jdsinger@dc.uba.ar

P. Zabala
e-mail: pzabala@dc.uba.ar

F. Alvelos et al. (eds.), *Advances in Optimization and Wildfire*, Lecture Notes in Operations Research, https://doi.org/10.1007/978-3-032-03108-2_14

1 Introduction

Forest wildfires represent an escalating threat, with both their frequency and intensity increasing over time [1]. The challenge of addressing this issue is compounded by the rate at which wildfires can spread [2–4]. In this context, the implementation of early detection systems capable of quickly alerting authorities is critical. Recognizing the urgency of this issue, governments around the world have explored various strategies for early wildfire detection. These include drone [5, 6] and satellite monitoring [7], predictive analysis [8], and aerial surveillance using helicopters and airplanes [9], and ground-based sensor networks [10]. In this paper we will focus on the latter approach. Ground-based sensors are a promising approach to early wildfire detection due to their low cost and little required maintenance.

One significant challenge in utilizing ground-based sensors for early wildfire detection is the limited availability of internet connectivity in rural areas. LoRa (Long Range) [11] is a wireless modulation technique derived from chirp spread spectrum (CSS) technology. It allows devices to communicate with each other across great distances, using very little power. LoRaWAN is the system architecture and communication protocol built on top of LoRa. LoRaWAN is particularly well-suited for applications requiring low power consumption and infrequent data transmission, such as environmental monitoring, smart agriculture, and wildfire detection systems [12–14]. Its ability to support a large number of devices in a network, coupled with its long-range capabilities, makes LoRaWAN an interesting solution for deploying extensive sensor networks in areas where traditional communication infrastructure is unavailable or impractical, such as forests.

In this paper, we present linear integer programming models to address the problem of optimally placing wildfire sensors. The paper is organized as follows: Section 1 introduces the necessary definitions for the problem description. Section 2 describes the MILP models developed for the problem. Section 3 presents the computational experiments conducted to evaluate the performance of these models. Section 4 provides a case study using the best-performing formulation from the evaluation. Section 5 concludes the paper with a summary of our findings and future research direction. Finally, we include an appendix with some theoretical results on the best-performing model.

2 Integer Programming Formulations

In this section, we formally define the sensor placement problem and present three alternative integer programming formulations.

2.1 Preliminary Definitions and Notation

In a LoRaWAN network, end devices will be responsible for sending data to gateways. Gateways act as intermediaries, forwarding this data to a network server, usually over an internet connection. Spreading factors (SF) determine the balance between data rate and communication range, while channels refer to the specific frequency bands used for transmitting data. In our scenario, wildfire sensors will act as our end devices.

Let be defined as P the set of *points of interest*, representing the locations we aim to protect, $S \subseteq P$ the set of points available to place a sensor, $S_p \subseteq S$ the set of points available for sensors that cover p, G the set of gateways, C the set of available channels, $k \in \mathbb{N}$ the number of sensors that can be used, n denotes the maximum number of sensors that can be assigned to a single gateway and m_{sg} the minimum spreading factor for sensor s to reach gateway g, or ∞ if sensor s cannot reach gateway g. In this context, we assume the goal is to use the minimum spreading factor necessary for a sensor to reach a given gateway, as this optimizes both the data transmission rate and energy consumption. We will define $priority_p$ to be the priority given to an interest point $p \in P$. For this paper, we assume that the priority range for a given location is limited to values from 1 to 3. Although this may vary across different scenarios, in this case we will consistently prioritize maximizing coverage over minimizing network interference.

This section describes the MILP models. Our goal is to determine:

- a subset of locations in which sensors are to be installed,
- a gateway and spreading factor associated with each sensor,
- a channel associated with each gateway.

The objective is to minimize a weighted combination of the number of uncovered points of interest, adjusted by their assigned priority, and the level of network interference, in such a way that the number of sensors does not exceed a prespecified number (given by budget constraints), and the number of sensors assigned to a gateway does not exceed a predefined number, given by hardware constraints. We consider *interference* between gateways g and h to occur if a sensor is assigned to gateway g but its signal also reaches gateway h, and both gateways operate on the same channel. Interference can occur in both directions, from h to g or from g to h, but it is not necessarily bidirectional.

2.2 Primary Model

We define the following binary decision variables for the model. The variable $l_p \in \{0, 1\}$ indicates whether the spot $p \in P$ is uncovered ($l_p = 1$) or covered ($l_p = 0$). The variable $i_{gh} \in \{0, 1\}$ represents the presence of *interference* between distinct gateways $g, h \in G$, with $i_{gh} = 1$ indicating there is a sensor s assigned to g whose signal also reaches h, and both g and h use the same channel and $i_{gh} = 0$ indicating otherwise. Similarly, $x_s \in \{0, 1\}$ specifies whether a sensor is located at point $s \in S$ ($x_s = 1$) or not ($x_s = 0$). The variable $z_{sg} \in \{0, 1\}$ denotes whether sensor $s \in S$ is assigned to gateway $g \in G$, with $z_{sg} = 1$ indicating an assignment. Finally, $u_{gc} \in \{0, 1\}$ indicates whether gateway $g \in G$ is using channel $c \in C$, with $u_{gc} = 1$ representing usage of the channel.

$$\text{Minimize:} \sum_{p \in P} \text{priority}(p) \cdot l_p + \frac{1}{|G|^2} \sum_{g \in G} \sum_{\substack{h \in G \\ g \neq h}} i_{gh} \tag{1}$$

Subject to:

$$\sum_{s \in S} x_s \leq k \tag{2}$$

$$\sum_{g \in G, m_{sg} \neq \infty} z_{sg} = x_s \quad \forall s \in S \tag{3}$$

$$\sum_{c \in C} u_{gc} = 1 \quad \forall g \in G \tag{4}$$

$$\sum_{s \in S} z_{sg} \leq n \quad \forall g \in G \tag{5}$$

$$l_p \geq 1 - \sum_{s \in S_p} x_s \quad \forall p \in P \tag{6}$$

$$z_{sg} + u_{gc} + u_{hc} - 2 \leq i_{gh} \forall c \in C, s \in S, g, h \in G,\ g \neq h, m_{sg} \geq m_{sh} \tag{7}$$

$$z_{sg} \in \{0, 1\} \quad \forall g \in G,\ \forall s \in S \tag{8}$$

$$l_p \in \{0, 1\} \quad \forall p \in P \tag{9}$$

$$x_s \in \{0, 1\} \quad \forall s \in S \tag{10}$$

$$u_{gc} \in \{0, 1\} \quad \forall g \in G,\ \forall c \in C \tag{11}$$

The objective function (1) seeks to minimize the number of areas that remain uncovered by any nearby sensor and at the same time, when possible and without affecting the number of covered sensors, minimize network interference. Constraints (2) limit the number of sensors used to at most k. Constraints (3) ensure that each sensor is assigned to exactly one gateway. Constraints (4) mandate that each gateway is allocated exactly one communication channel. Constraints (5) impose a restriction that no gateway can have more than n sensors assigned to it. Constraints (6) stipulate that a location can only be considered covered if a sensor capable of covering it is deployed. Constraints (7) ensure that there is interference between gateways when a sensor is assigned to one gateway but reaches another gateway that uses the same channel. We show in the appendix that these last group of constraints are facet-defining. Finally, constraints (8–11) define the domain for the decision variables. For simplicity, we will refer to this as the *primary model* throughout the paper.

We will introduce two additional integer programming models to compare alternative approaches for modeling channel assignment.

2.3 Representatives model

The variables in this model are defined identically to those in the previous model, except for variables u_{gc}, which are excluded. Additionally, we introduce the variables $v_{hg} \in \{0, 1\}$, which indicate whether gateway $g \in G$ shares the same channel as gateway h. When $v_{hg} = 1$, we will say that gateway $h \in G$ *represents* the channel of g.

$$\text{Minimize: } \sum_{p \in P} \text{priority}(p) \cdot l_p + \frac{1}{|G|^2} \sum_{g \in G} \sum_{\substack{h \in G \\ g \neq h}} i_{gh} \tag{12}$$

Subject to:

$$\sum_{s \in S} x_s \leq k \tag{13}$$

$$\sum_{g \in G, m_{sg} \neq \infty} z_{sg} = x_s \quad \forall s \in S \tag{14}$$

$$\sum_{s \in S} z_{sg} \leq n \quad \forall g \in G \tag{15}$$

$$l_p \geq 1 - \sum_{s \in S_p} x_s \quad \forall p \in P \tag{16}$$

$$\sum_{h \in G} v_{hg} = 1 \quad \forall g \in G \tag{17}$$

$$v_{gh} \leq v_{gg} \quad \forall g, h \in G \tag{18}$$

$$\sum_{g \in G} v_{gg} \leq |C| \tag{19}$$

$$z_{sg} + v_{jg} + v_{jh} - 2 \leq i_{gh} \forall g,\ h,\ j \in G,\ \forall s \in S,\ m_{sg} \geq m_{sh} \tag{20}$$

$$z_{sg} \in \{0, 1\} \quad \forall g \in G, \forall s \in S \tag{21}$$

$$l_p \in \{0, 1\} \quad \forall p \in P \tag{22}$$

$$x_s \in \{0, 1\} \quad \forall s \in S \tag{23}$$

$$v_{gh} \in \{0, 1\} \quad \forall g \in G, \forall h \in G \tag{24}$$

The objective function (12) is identical to the objective function (1). Constraints (13), (14), (15), (16) are the same as in the previous model and constraints (21)–(24) specify the domain of the variables. Constraints (17) ensure that each gateway is assigned exactly one channel. Constraints (19) limit the total number of channels used to at most $|C|$. Constraints (18) ensure that if a gateway represents its channel, it must also represent itself, while constraints (20) enforce the validity of interference conditions. Restrictions (21) to (24) define the domain of the variables.

2.4 Partial Order Channel Model

This model uses the same variable definitions as the primary model, except for the variables u_{gc}, which are not included. Additionally, we introduce the variables $L_{gc} \in \{0, 1\}$, which specify whether gateway g uses a channel $c' < c$ (assuming the channels are ordered), and $U_{gc} \in \{0, 1\}$, which indicate whether gateway g uses a channel $c' > c$.

$$\text{Minimize: } \sum_{p \in P} \text{priority}(p) \cdot l_p + \frac{1}{|G|^2} \sum_{g \in G} \sum_{\substack{h \in G \\ g \neq h}} i_{gh} \tag{25}$$

Subject to:

$$\sum_{s \in S} x_s \leq k \tag{26}$$

$$\sum_{g \in G, m_{sg} \neq \infty} z_{sg} = x_s \quad \forall s \in S \tag{27}$$

$$\sum_{s \in S} z_{sg} \leq n \quad \forall g \in G \tag{28}$$

$$l_p \geq 1 - \sum_{s \in S_p} x_s \quad \forall p \in P \tag{29}$$

$$L_{g1} = U_{g|C|} = 0 \quad \forall g \in G \tag{30}$$

$$U_{g(c-1)} - U_{gc} \geq 0 \quad \forall g \in G,\ c \in C \tag{31}$$

$$L_{g(c+1)} - L_{gc} \geq 0 \quad \forall g \in G,\ c \in C \tag{32}$$

$$L_{g(c+1)} + U_{gc} = 1 \quad \forall g \in G,\ c \in C,\ c \leq |C| - 1 \tag{33}$$

$$z_{sg} + \begin{cases} U_{g(c+1)} + L_{g(c-1)} + U_{h(c-1)} + L_{g(c+1)} - 4 \leq i_{gh} \text{ if } 1 < c < |C| \\ U_{g(c+1)} + L_{g(c+1)} - 2 \leq i_{gh} \text{ if } c = 1 \text{ and } |C| > 1 \\ L_{g(c-1)} + U_{h(c-1)} - 2 \leq i_{gh} \text{ if } c = |C| \text{ and } c > 1 \\ 0 \leq i_{gh} \text{ if } c = 1 = |C| \end{cases} \tag{34}$$

$$\forall g, h \in G,\ g \neq h,\ \forall c \in C,\ \forall s \in S,\ m_{sg} \geq m_{sg'}$$

$$\sum_{j \in S} z_{jg} \leq n \quad \forall g \in G \tag{35}$$

$$i_{gh} \in \{0, 1\} \quad \forall g, h \in G,\ g \neq h \tag{36}$$

$$z_{sg} \in \{0, 1\} \quad \forall s \in S,\ g \in G \tag{37}$$

$$L_{gc} \in \{0, 1\} \quad \forall g \in G, c \in C \tag{38}$$

$$U_{gc} \in \{0, 1\} \quad \forall g \in G, c \in C \tag{39}$$

$$l_p \in \{0, 1\} \quad \forall p \in P \tag{40}$$

$$x_s \in \{0, 1\} \quad \forall s \in S \tag{41}$$

The objective function (25) is identical to the objective function (1). Constraints (26), (27), (28), (29) and (35) are the same as in the previous models. Constraint (30) ensures that each gateway is assigned a valid channel between 1 and $|C|$. Constraints (31) to (33) ensure the coherence of channel ordering for variables U_{gc} and L_{gc} for each $g \in G$ and $c \in C$. Constraints (34) assert the validity of the interference. Restrictions (36) to (41) define the domains of the variables.

3 Computational Experiments

In this section, we present the results of computational experiments conducted using the integer programming models introduced in the previous section. Throughout this paper, we used Python along with the Gurobi Optimizer (version 12.0.0, build v12.0.0rc1) with its default configuration settings, which ran on an Apple M1 Pro CPU, with 8 physical cores, 8 logical processors, and using up to 8 threads. A time limit of 30 minutes was applied, and the `MIPGap` parameter was set to 0.01%.

The instances were randomly generated following this procedure: first, we predefined the values for the number of interest points ($|P|$), sensor points ($|S|$), gateways ($|G|$), communication channels ($|C|$), and the sensor limit available (k). We represented the interest points with numbers going from 1 to $|P|$. Subsequently, $|S|$ locations were randomly selected from the set of interest points, identifying the potential locations for sensor placement. We considered the notion of proximity between interest points according to their index. For each potential sensor location we randomly choose between 1 and 7 the radius of interest points covered by that location, that is, the nearest indexes of interest points covered by it. Lastly, we randomly created the priority score between 1 and 3, with 50% probability of having the lowest priority and 20% probability having the highest. In all cases we assumed $n = 100$.

Table 1 shows the results of the solver across different instances and models. The first five columns provide data about each instance, while the subsequent three groups of columns display the performance of the different models. The column "Time(s)" reports the running time to optimality. The column "Nodes" reports the number of nodes in the enumeration tree at the termination of the branch and cut procedure. Table 2 shows the size of the instances for each model, including the number of variables and constraints.

4 Case Study

In this section, we present a practical example of the primary model applied to Pereyra Iraola Park, a provincial park located in Buenos Aires Province, Argentina. This case study serves as a real-world application of the proposed integer programming model, demonstrating its potential to address wildfire detection challenges in ecologically sensitive areas.

Spanning 10,246 hectares, Pereyra Iraola Park is one of the largest urban parks in the world. Its rich biodiversity makes it a crucial site for conservation efforts and a popular destination for recreational activities. However, the increasing frequency of wildfires poses a significant threat to its ecosystem, emphasizing the need for

Table 1 Comparison of computational performance across the primary model and its variants for random instances

Instance					Primary model		Representatives model		Partial order model	
$\|P\|$	$\|S\|$	$\|G\|$	k	$\|C\|$	Time(s)	Nodes	Time(s)	Nodes	Time(s)	Nodes
5500	4000	3	300	1	1.5	1	1.7	1	1.4	1
7500	4000	8	600	4	7.3	1	20.5	1	9.5	1
7500	2000	8	600	4	2.6	1	4.3	1	3.4	1
10500	6000	8	300	4	12.2	1	16.5	1	14.8	1
10500	6000	8	600	1	26.0	1	80.7	1	25.8	1
9500	7000	9	600	2	13.0	1	107.3	1	8.7	1
9500	2000	4	300	4	0.34	1	0.15	1	0.43	1
13500	7000	9	900	3	26.0	1	82.1	1	32.3	1
13500	2000	4	300	2	0.3	1	0.5	1	0.3	1
9500	9500	9	900	3	53.3	1	1457.0	1	111.4	1
5500	5500	7	300	4	5.5	1	11.1	1	8.9	1
5500	5500	4	300	4	1.6	1	0.7	1	2.8	1
7500	7500	7	600	2	30.7	1	465.6	136	17.0	84
7500	7500	7	600	4	26.7	1	198.7	1	15.0	1
3500	3500	9	300	2	23.3	103	258.8	15	36.3	95
3500	3500	9	400	2	9.0	1	339.2	15	9.3	1
3500	3500	9	500	2	2.2	1	220.6	23	3.0	1
7500	7500	7	500	2	6.9	1	54.1	1	9.1	1

an effective early detection system. Figure 1 shows an outline of the park, with the black-shaded regions representing private property within the park boundaries, where sensor placement is prohibited.

We analyzed a configuration consisting of three strategically placed gateways within the park. The central gateway was located near the ranger station, one additional gateway was placed in the southern region close to the park's trails, and the third was positioned in the northern section within the native vegetation area. The park will be divided into a finite set of interest points, each separated by 200 meters.

Different levels of coverage were considered in various areas of the park, determined by the type of vegetation present in each region. Specifying the coverage of a sensor is a complex task that depends on many factors, including specifics of the sensor we will use, characteristics of the terrain, and the stage at which we expect to catch a wildfire [11]. This case study aims to estimate coverage levels, with the intention of refining this step in future work.

Additionally, higher priority will be assigned to areas that include trails or are accessible to the general public, followed by secondary priority for regions hosting native protected vegetation. A limit of 300 sensors overall and a limit of 100 sensors for each gateway were used in the analysis.

Table 2 Size of the instances used in Table 1. The table reports the number of variables and constraints for each model

Instance					Primary model		Representatives model		Partial order model	
$\|P\|$	$\|S\|$	$\|G\|$	k	$\|C\|$	#(Variables)	#(Constraints)	#(Variables)	#(Constraints)	#(Variables)	#(Constraints)
5500	4000	3	300	1	21512	23171	21518	50509	21515	23174
7500	4000	8	600	4	43596	523313	43628	1035174	43628	523393
7500	2000	8	600	4	25596	265965	25628	522478	25628	266045
10500	6000	8	300	4	64596	784617	64628	1552782	64628	784697
10500	6000	8	600	1	64572	208720	64628	1554206	64580	208728
9500	7000	9	600	2	79599	592165	79662	2607008	79617	592201
9500	2000	4	300	4	19532	66565	19532	66582	19548	66605
13500	7000	9	900	3	83608	884339	83662	2612061	83635	884402
13500	2000	4	300	2	23524	42927	23532	70362	23532	42943
9500	9500	9	900	3	104608	1191809	104662	3537471	104635	1191872
5500	5500	7	300	4	49577	538999	49598	935037	49605	539069
5500	5500	4	300	4	33032	162021	33032	162038	33048	162061
7500	7500	7	600	2	67563	375197	67598	1275702	67577	375225
7500	7500	7	600	4	67577	735171	67598	1275338	67605	735241
3500	3500	9	300	2	38599	294695	38662	1301643	38617	294731
3500	3500	9	400	2	38599	295103	38662	1303479	38617	295139
3500	3500	9	500	2	38599	294609	38662	1301256	38617	294645
7500	7500	7	500	2	67563	375165	67598	1275590	67577	375193

Fig. 1 Pereyra Iraola Park in Buenos Aires Province

We ran the primary model with a time limit of 120 seconds. We obtained a model with 24, 651 constraints and 14, 472 variables. The final objective function value was 385.1, with a gap of 2.34%. Figure 2 shows the final result of sensor placement, with the white dots representing the locations where sensors should be placed.

We observe that some regions exhibit larger gaps between sensors, corresponding to areas with low vegetation, therefore we assumed to have more coverage. The lower central region, which has the highest priority, contains the highest density of sensors. Additionally, sensors tend to cluster around the boundaries of the forbidden areas.

This study demonstrated the ability of the MILP model to address the Sensor Placement In LoraWAN Networks problem in a real-world context, such as Pereyra Iraola Park. This highlights its effectiveness in planning early detection systems for critical conservation areas. Further refinement of coverage estimation and adjustments for additional environmental or sensor-specific factors could enhance the model's precision and applicability in similar contexts.

5 Concluding remarks

In this work, we have presented an integer programming formulation for the sensor placement problem in LoRaWAN networks, a topic of particular interest due to the increasing frequency of wildfires in recent years. Moreover, LoRaWAN technology emerges as a promising solution for addressing challenges in areas with limited

Fig. 2 Final sensor placement result: The white circles indicate the optimal locations for the 300 sensors. The markers indicate the locations of the 3 gateways

internet connectivity, lack of power infrastructure, and the need for devices capable of withstanding natural conditions such as rain or heat. We compared our model against two variants using randomly generated instances, and evaluated its performance using a real-life case study focused on Pereyra Iraola park. Our results indicate that the complexity of the problem is strongly influenced by what we define as network interference, which can be affected by the number of gateways and available channels. With the current model, increasing the number of channels significantly reduces the resolution time of the problem, even though interference impacts the objective function with a penalty of only one.

When analyzing our primary model, we observed that most solutions to the random instances were obtained at the root node, with the dual bound being relatively close to the final solution. This observation suggests that it may be beneficial to explore heuristic approaches to further reduce execution times. Additionally, this work can be extended by considering the various types of interference that LoRaWAN networks may experience, which can depend on the use case and other environmental conditions.

Appendix

In this section, we present some theoretical results related to the polytope induced by the primary formulation. We begin by determining the dimension of the polytope and proceed to prove the facet-defining property of certain inequalities in the model.

Definition 1 We define $\mathcal{P}(P, S, G, C) \subset \mathbb{R}^{|V||C|+|\mathcal{F}|}$ to be the convex hull of all points $(x, z, u, l, i) \in \mathbb{R}^{|S|+|G||S|+|C||G|+|P|+|G|(|G|-1)}$ satisfying constraints (2)–(11).

Note that the polytope does not depend on k or n since we assume these constants to be fixed values. Throughout this section we will assume reasonable hypotheses on the input data. We state these assumptions formally for future reference.

Assumption 1 We assume that for every point of interest $p \in P$, there exists a sensor point $s \in S$ such that p can be covered by s. We will call this set of points S_p.

If Assumption 1 does not hold, i.e., there exists a point $p \in P$ such that it cannot be covered by any sensor point s, then we will simply fix $l_p = 1$.

Assumption 2 For every point $s \in S$, there exists a gateway $g \in G$ such that $m_{sg} < \infty$. This means every sensor can reach at least one gateway.

Assumptions 1 and 2 will be necessary for characterizing the dimension of $\mathcal{P}(P, S, G, C)$ and in the facetness proof. We denote by $\hat{0}$ the all-zeros vector with the appropriate dimension.

Proposition 1 The dimension of $\mathcal{P}(P, S, G, C)$ is $|P| + |G|(|G| - 2) + |S||G| + |G||C|$.

Proof We aim to demonstrate that there are no additional linearly independent equalities beyond those explicitly defined in the model. The equalities in the model are linearly independent, as they are defined over distinct sets of variables. Given that these equalities constitute a total of $|S| + |G|$, proving the absence of any further linearly independent equalities would thus establish the proposition.

To achieve this, consider a generic equality of the form $\alpha^\intercal l + \beta^\intercal i + \gamma^\intercal x + \delta^\intercal z + \omega^\intercal u = \mu$ which is satisfied by all points within the polytope. We will show that this equality must necessarily be a linear combination of the equalities already defined in the model.

Claim 1: $\alpha_p = 0 \; \forall p \in P$: Let $p \in P$ fixed. Let $s \in S$ be such that $s \in S_p$, i.e., s covers point p. Take a feasible solution w_1 such that $x_s = 1$ and $x_{s'} = 0 \; \forall s' \neq s$. We will take $g \in G$ with $m_{sg} < \infty$, and we will fix the vector z in such a way that $z_{s'g'} = 0$ for all $(s', g') \in (S, G) \setminus \{(s, g)\}$, and $z_{sg} = 1$. We will also fix $i_{gh} = 1$ $\forall g, h \in G, g \neq h$, and $u_{g1} = 1 \; \forall g \in G$, $u_{gc} = 0 \; \forall g \in G, \; \forall c \in C, c > 1$. Lastly, we will fix $l_q = 1 \; \forall q \in P$. Notice this solution is valid since it satisfies the model constraints. We then modify the solution by setting $l_p = 0$. Both solutions satisfy the model constraints and hence we prove the claim.

Claim 2: $\beta_{gh} = 0 \ \forall g \neq h$: Let $g', h' \in G$ fixed, with $g' \neq h'$. Take a feasible solution with no sensors placed, such that every gateway is assigned the same channel $c \in C$. We will fix $z_{sg} = 0 \ \forall s \in S, \forall g \in G$, $l_p = 1 \ \forall p \in P$ and $i_{gh} = 1 \ \forall g, h \in G$ distinct.

Without breaking any restriction, we can modify this solution setting $i_{g'h'} = 0$.

Claim 3: $\omega_{gc} = \omega_{gc'}$: Fixing $g \in G$, $c, c' \in C$, this claim can be proven using a similar approach as the previous one, forming first a feasible solution with no sensors and $u_{gc} = 1$, and then modifying it only by setting $u_{gc'} = 1$.

Claim 4: $\gamma_s = -\delta_{sg} \ \ \forall s \in S, \forall g \in G$, with $m_{sg} < \infty$: We will fix $s \in S$, $g \in G$ such that $m_{sg} < \infty$. We can then build a solution w_1, with $x_{s'} = 0 \ \forall s \in S$, $z_{s'g'} = 0 \ \forall s' \in S, g' \in G$. Lastly, we can fix $l_p = 1 \forall p \in P$ and $i_{gh} = 1 \ \forall g, h \in G$ distinct. Let us form a second solution w_2, modifying w_1 by setting $x_s = z_{sg} = 1$. Since both are feasible solutions, we have $\gamma_s + \delta_{sg} = 0$.

Claim 5: $\delta_{sg'} = \delta_{sg} \ \ \forall s \in S, \forall g \in G, m_{sg} < \infty$: This claim can be inferred from the previous one.

Claim 6: $\delta_{s'g'} = 0 \ \ \forall s' \in S, \forall g' \in G, m_{s'g'} = \infty$: This claim is trivial since those variables don't appear on any of the restrictions. Therefore, we can demonstrate that the equality is a combination of the model's equalities, thereby proving the proposition. □

We will now prove restriction 7 is facet-defining.

Proposition 2 The interference inequalities

$$z_{sg} + u_{gc} + u_{hc} - 2 \leq i_{gh}$$

define a facet for all $g, h \in G$ distinct, and for all $s \in S$ if assumptions 1 and 2 are satisfied.

Proof Let us assume a generic equality $\alpha^\top l + \beta^\top i + \gamma^\top x + \delta^\top z + \omega^\top u = \mu$ within the face induced by the equality $z_{sg} + u_{gc} + u_{hc} - 2 = i_{gh}$. In order to prove the proposition, we need to show that this equality is a linear combination of the model's equalities and the interference equality.

Claim 1: $\boldsymbol{\alpha_p = 0} \ \forall p \in P$: Fix $p \in P, d \in C, g_1 \in G, s_p \in S_p$ such that $m_{s_p g_1} < \infty$. If $g_1 = g$ and $s_p = s$, set $u_{gc} = u_{hc} = z_{sg} = i_{gh} = 1$. We will also fix the values: $l_q = 1 \ \forall q \in P$, $i_{jk} = 1 \ \forall j, k \in G$, $x_{s'} = 0 \ \forall s' \in S \setminus \{s_p\}$, $x_{s_p} = 1$, $z_{s_p g_1} = 1$, and $z_{s'g'} = 0 \ \forall (s', g') \in (S, G) \setminus \{(s_p, g_1)\}$, $u_{gc} = u_{hc} = 1$, and $u_{g'd} = 1 \ \forall g' \in G \setminus \{g, h\}$.

If $g_1 = g$, $s_p \neq s$ and $m_{s_p g} < m_{s_p h}$ we can fix the values: $z_{sg} = i_{gh} = 0$, and $u_{gc} = 1 = u_{hc}$. We also fix the values: $l_q = 1 \ \forall q \in P$, $i_{jk} = 1 \ \forall j, k \in G$ such that $(j, k) \neq (g, h)$, $i_{gh} = 0$, $x_{s'} = 0 \ \forall s' \in S, s' \neq s_p$, $x_{s_p} = 1$, $z_{s_p g_1} = 1$, $z_{s'g'} = 0 \ \forall s' \in S, g' \in G$ such that $(s', g') \neq (s_p, g_1)$, $u_{gc} = u_{hc} = 1$, and $u_{g'd} = 1 \ \forall g' \in G, g' \neq g, \ g' \neq h$.

If $g_1 = g$, $s_p \neq s$ and $m_{s_p g} \geq m_{s_p h}$ we can fix the values: $z_{sg} = i_{gh} = 1 = x_s$, and $u_{gc} = 1 = u_{hc}$. We can also fix the values: $l_q = 1 \ \forall q \in P$, $i_{jk} = 1 \ \forall j, k \in G$, $x_{s'} = 0 \ \forall s' \in S, s \neq s' \neq s_p$, $x_{s_p} = 1$, $z_{s_p g} = 1$, $z_{sg} = 0 \ \forall s \in S, g \in G$ such that $(s, g) \neq (s_p, g_1)$, $u_{gc} = u_{hc} = 1$, and $u_{g'd} = 1 \ \forall g' \in G, g' \neq g, \ g' \neq h$.

If $g_1 \neq g$, we will have the following values: $l_q = 1\ \forall q \in P$, $i_{jk} = 1\ \forall j,k \in G$ such that $(j,k) \neq (g,h)$, $i_{gh} = 0$, $x_s = 0\ \forall s' \in S, s' \neq s_p$, $x_{s_p} = 1$, $z_{s_p g_1} = 1$, $z_{s'g'} = 0\ \forall s' \in S, g' \in G$ such that $(s', g') \neq (s_p, g_1)$, $u_{gc} = u_{hc} = 1$, $u_{g'd} = 1\ \forall g' \in G, g' \neq g,\ g' \neq h$.

In every case, we have a valid solution w_1, that satisfies the interference equality. We can modify this solution to form w_2, with the same values but having $l_p = 0$. Given that the two solutions are valid, we have the statement proven.

Claim 2: $\boldsymbol{\beta_{g_1h_1} = 0}$ $\forall g_1, h_1 \in G$ distinct, with $(g_1, h_1) \neq (g, h)$. To prove this statement, let $g_1 \in G$, $h_1 \in G$ and $d \in C$. We can create a solution w_1 with the following values: $l_q = 1\ \forall q \in P$, $i_{jk} = 0\ \forall j,k \in G$, $x_{s'} = 0\ \forall s' \in S$, $z_{s'g'} = 0\ \forall s' \in S, g' \in G$, $u_{gc} = u_{hc} = 1$, $u_{g'd} = 1\ \forall g' \in G, g' \neq g,\ g' \neq h$. This solution is valid and satisfies the interference equality. Lastly, we take a second solution w_2 with the same values but we fix $i_{g_1h_1} = 1$.

Claim 3: $\gamma_{s_1} = -\delta_{s_1g_1}$ $\forall s_1 \in S, \forall g_1 \in G,\ m_{s_1g_1} < \infty$ and $(s_1, g_1) \neq (s, g)$. Let $s_1 \in S, g_1 \in G$ and $d \in C$.

- If $s = s_1$ and $g_1 \neq g$: Let us construct w_1 with the values $l_q = 1\ \forall q \in P$, $i_{jk} = 1\ \forall j,k \in G$ such that $(j,k) \neq (g,h)$, $i_{gh} = 0, x_{s'} = 0\ \forall s' \in S \setminus \{s_1\}, x_{s_1} = 1$, $z_{s'g'} = 0\ \forall s' \in S, g' \in G$ such that $(s', g') \neq (s_1, g_1)$, $z_{s_1g_1} = 1$, $u_{gc} = u_{hc} = 1$, $u_{g'd} = 1\ \forall g' \in G, g' \neq g,\ g' \neq h$.
- If $g \neq g_1$ or $m_{s_1g} < m_{s_1h}$: We can create a solution w_1 with the following values: $l_q = 1\ \forall q \in P$, $i_{jk} = 1\ \forall j,k \in G$ such that $(j,k) \neq (g,h)$, $i_{gh} = 0$, $x_{s'} = 0\ \forall s' \in S \setminus \{s_1\}$, $x_{s_1} = 1$, $z_{s'g'} = 0\ \forall s' \in S, g' \in G$ such that $(s', g') \neq (s_1, g_1)$, $z_{s_1g_1} = 1$, $u_{gc} = u_{hc} = 1$, $u_{g'd} = 1\ \forall g' \in G, g' \neq g,\ g' \neq h$.
- If $g = g_1$ and $m_{s_1g} \geq m_{s_1h}$: We will create a solution w_1 with the following values: $l_q = 1\ \forall q \in P$, $i_{jk} = 1\ \forall j,k \in G$, $x_{s'} = 0\ \forall s' \in S \setminus \{s_1, s\}$, $x_{s_1} = 1 = x_s$, $z_{s'g'} = 0\ \forall s' \in S, g' \in G$ such that $(s,g) \neq (s', g') \neq (s_1, g_1)$, $z_{s_1g_1} = 1 = z_{sg}$, $u_{gc} = u_{hc} = 1$, $u_{g'd} = 1\ \forall g' \in G, g' \neq g,\ g' \neq h$

Lastly, we will create w_2 by removing sensor s_1, i.e. setting $x_{s_1} = 0 = z_{s_1g_1}$. Both are valid solutions that live in the equality, hence we obtain $\gamma_s + \delta_{sg} = 0$.

Claim 4: $\gamma_s = -\delta_{sg} - \beta_{gh}$. Fix $d \in C$. Let us construct a solution w_1 with: $l_q = 1\ \forall q \in P$, $i_{jk} = 1\ \forall j,k \in G$, $x_{s'} = 0\ \forall s' \in S \setminus \{s\}$, $x_s = 1$, $z_{s'g'} = 0\ \forall s' \in S, g' \in G$ such that $(s', g') \neq (s, g)$, $z_{sg} = 1$, $u_{gc} = u_{hc} = 1$, $u_{g'd} = 1\ \forall g' \in G, g' \neq g,\ g' \neq h$.

Then we can create w_2, identical to w_1 but with: $i_{gh} = 0$, $x_s = 0$ and $z_{sg} = 0$.

Since both solutions are valid and satisfy our equality, we have $\gamma_s + \delta_{sg} + \beta_{gh} = 0$.

Claim 5: $\omega_{g_1d} = \omega_{g_1d'}$ $\forall g_1 \in G, d, d' \in C, g_1 \neq \{g, h\}$. Fix $g_1 \in G \setminus \{g, h\}$. Fix $d, d' \in C$. If $d = d'$ the claim is trivial, so we assume $d' \neq d$. Let us create a solution w_1 with the following values: $l_q = 1\ \forall q \in P$, $i_{jk} = 1\ \forall j,k \in G$, $x_{s'} = 0\ \forall s' \in S \setminus \{s\}$, $x_s = 1$, $z_{s'g'} = 0\ \forall s' \in S, g' \in G$, $u_{gc} = u_{hc} = 1$, $u_{g'd} = 1\ \forall g' \in G, g' \neq g,\ g' \neq h$.

Next, we can create w_2 identically to w_1, but fixing $u_{g_1d'} = 1$, $u_{g_1d} = 0$. We can prove the claim using w_1 and w_2.

Claim 6: $\omega_{gd} = \omega_{gc} + \beta_{gh}$ $\forall d \in C \setminus \{c\}$. Let us fix $d \in C \setminus \{c\}$. We form a solution w_1 with the values $l_q = 1\ \forall q \in P$, $i_{jk} = 1\ \forall j,k \in G$, $x_{s'} = 0\ \forall s' \in S \setminus \{s\}$,

$x_s = 1$, $z_{s'g'} = 0\ \forall s' \in S, g' \in G$ with $(s', g') \neq (s, g)$, $z_{sg} = 1$, $u_{hc} = 1$, $u_{gc} = 1$, $u_{g'd} = 1\ \forall g' \in G, g' \neq g,\ g' \neq h$.

We then form a second solution w_2 with the values $l_q = 1\ \forall q \in P$, $i_{jk} = 1\ \forall j, k \in G\ (j, k) \neq (g, h)$, $i_{gh} = 0$, $x_{s'} = 0\ \forall s' \in S \setminus \{s\}$, $x_s = 1$, $z_{s'g'} = 0\ \forall s' \in S, g' \in G$, $u_{hc} = 1$, $u_{gd} = 1$, $u_{g'd} = 1\ \forall g' \in G, g' \neq g,\ g' \neq h$.

Claim 7: $\omega_{hd} = \omega_{hc} + \beta_{gh}$. The proof is analogous to the previous one and is therefore omitted for brevity.

Claim 8: $\delta_{s'g_1} = \delta_{s'g}\ \forall s' \neq s,\ g_1 \in G,\ m_{s'g_1} + m_{s'g} < \infty$. This result can be directly inferred from Claim 3.

Claim 9: $\delta_{s'g_1} = 0\ \forall s' \in S,\ g_1 \in G$ such that $m_{sg} = \infty$. This claim is trivial.

We can thereby yield the desired result. □

References

1. Weber, K.T., Yadav, R.: Spatiotemporal trends in wildfires across the Western United States (1950–2019). Remote Sens. **12**(18), 2959 (2020). https://doi.org/10.3390/rs12182959
2. Storey, M., Bedward, M., Price, O., Bradstock, R., Sharples, J.: Derivation of a Bayesian fire spread model using large-scale wildfire observations (2021). https://doi.org/10.1016/j.envsoft.2021.105127
3. Agranat, V., Perminov, V.: Mathematical modeling of wildland fire initiation and spread. Environ. Model. Softw. **125**, 104640 (2020). https://doi.org/10.1016/j.envsoft.2020.104640
4. Jiang, W., Wang, F., Su, G., Li, X., Wang, G., Zheng, X., Wang, T., Meng, Q.: Modeling wildfire spread with an irregular graph network. Fire **5**(6), 185 (2022). https://doi.org/10.3390/fire5060185
5. Abdusalomov, A., Umirzakova, S., Bakhtiyor Shukhratovich, M., Mukhiddinov, M., Kakhorov, A., Buriboev, A., Jeon, H.S.: Drone-based wildfire detection with multi-sensor integration. Remote Sens. **16**(24), 4651 (2024). https://doi.org/10.3390/rs16244651
6. Kinaneva, D., Hristov, G., Raychev, J., Zahariev, P.: Early forest fire detection using drones and artificial intelligence. In: 2019 42nd International Convention on Information and Communication Technology, Electronics and Microelectronics (MIPRO), Opatija, Croatia, pp. 1060–1065 (2019). https://doi.org/10.23919/MIPRO.2019.8756696
7. Rashkovetsky, D., Mauracher, F., Langer, M., Schmitt, M.: Wildfire detection from multisensor satellite imagery using deep semantic segmentation. IEEE J. Sel. Top. Appl. Earth Observations Remote Sens. **14**, 7001–7016 (2021). https://doi.org/10.1109/JSTARS.2021.3093625
8. Xu, Z., Li, J., Cheng, S., Rui, X., Zhao, Y., He, H., Xu, L.: Wildfire risk prediction: A review. arXiv preprint, arXiv:2405.01607, 2024. Available at: https://arxiv.org/abs/2405.01607
9. Marchi, E., Neri, F., Tesi, E., Fabiano, F., Montorselli, N.: Analysis of helicopter activities in forest fire-fighting. Croat. J. For. Eng. **35**, 234–243 (2014)
10. Chan, C.C., Alvi, S., Zhou, X., Durrani, S., Wilson, N., Yebra, M.: A survey on IoT ground sensing systems for early wildfire detection: Technologies, challenges and opportunities. IEEE Access, 1-1 (2024). 10.1109/ACCESS.2024.3501336
11. LoRa Alliance: LoRa Alliance: The leading industry association for the LoRaWAN standard. [Online]. Available via DIALOG https://lora-alliance.org/. Accessed 15 Nov 2024
12. Wang, Y., Huang, Y., Song, C.: A new smart sensing system using LoRaWAN for environmental monitoring. In: 2019 Computing, Communications and IoT Applications (ComComAp), pp. 347-351. Shenzhen, China (2019). https://doi.org/10.1109/ComComAp46287.2019.9018829

13. Badreddine, M., Bourennane, E.-B., Boucherkha, S., Chikhi, S.: A study of LoRaWAN protocol performance for IoT applications in smart agriculture. Comput. Commun. **164**, 148–157 (2020). ISSN 0140-3664. https://doi.org/10.1016/j.comcom.2020.10.009
14. Mukhia, R., Sriporamanont, K., Sinthupinyo, S.: Performance evaluation of LoRaWAN forest fire monitoring network in the wild. In: Proceedings of the 18th Asian Internet Engineering Conference (AINTEC '23), Article No. 2, pp. 1–8, December 2023. https://doi.org/10.1145/3630590.3630602.

Optimizing Autonomous Unmanned Aircraft System Deployment Locations for Enhanced Wildfire Detection and Monitoring

Sascha Emanuel Zell and Armin Fügenschuh

Abstract The increase in the frequency and severity of wildfires as a symptom of climate change requires innovative methods of wildfire fighting. For this reason, we propose a framework for an autonomous Unmanned Aircraft System (UAS), consisting of a fleet of Unmanned Aerial Vehicles (UAVs) stored in purpose-built hangars. The intention is to deploy the UAS as the first responder to an alarm and have the sensor-equipped UAVs monitor the target area even before other standard firefighting vehicles have arrived. The focus of this paper is primarily on the development and application of a location-allocation optimization Mixed-Integer Linear Programming (MILP) model that selects different UAV and hangar types and locates them, with the objective of approaching the target area as quickly as possible while guaranteeing a certain monitoring time at the scene. The model is applied to a large, sparsely populated, rural operational area, around a third of which consists of forest in the South of Brandenburg, Germany. The spatial demand is measured through an easily reproducible and transferable open data approach. Finally, several instances with different fixed numbers of hangars and UAVs to be set up are solved by the commercial state-of-the-art solver CPLEX and analyzed for their computation time.

Keywords Mixed-Integer linear programming · Unmanned aircraft systems · Unmanned aerial vehicles · Autonomous wildfire monitoring · Location-allocation optimization

---

S. E. Zell (✉) · A. Fügenschuh
Brandenburg University of Technology, Cottbus, Germany
e-mail: Sascha.Zell@b-tu.de

A. Fügenschuh
e-mail: fuegensc@b-tu.de

F. Alvelos et al. (eds.), *Advances in Optimization and Wildfire*, Lecture Notes in Operations Research, https://doi.org/10.1007/978-3-032-03108-2_15

1 Introduction

Wildfires are a major challenge for emergency services as they can destroy large forest stands and thus biodiversity, pose a threat to life and require a large amount of resources to fight. Climate change and the associated temperature extremes and periods of heat and drought in the summer months are contributing to an increasing risk of wildfires. Due to limited human resources, innovative technologies and ideas must be implemented to support the emergency services. Many technologies and optimization approaches have recently been presented for this purpose, and this is where we contribute by promoting the use of Unmanned Aerial Vehicles (UAVs) for fighting wildfires. UAVs are highly suitable for dangerous tasks, as they do not bring people into danger zones, as do firetrucks or helicopters. In addition, UAVs and especially so-called Micro Aerial Vehicles (MAVs) are of keen interest because of their small size, excellent maneuverability including the ability to hover and their ability to cover large areas quickly [1]. There is also a wide range of different sensors, e.g. LiDAR (Light Detection and Ranging), Radar (Radio Detection and Ranging), Sonar or Image sensors, that can be mounted on a UAV and used for suitable applications such as fire source and smoke detection.

In the following, we present an idea that was developed in close cooperation with a local control center in Brandenburg, Germany, the *Leitstelle Lausitz*. The department is responsible for coordinating emergency medical service, firefighting and disaster operations. The initiative was triggered by the increasing number of hot and dry periods in summer in Central Europe, with the Lusatia region in particular being considered one of the driest regions in Germany. Furthermore, more than 70 % of the trees in Brandenburg's forests are pine trees (*Pinus sylvestris L.*) which dry out particularly quickly and thus increase the risk of wildfire outbreak and spread [2]. One proposed solution is to use MAVs as first responders after receiving a wildfire report. It is expected that they will be the first to arrive at the scene, long before any other rescue services. The main reasons for this are the high speed of MAVs and the fact that they are not tied to road networks and, unlike fire trucks or helicopters, do not need to be equipped with a crew and are ready for action after a very short lead time. After arriving in the target area, the UAVs can then monitor and search the area together and use sensors to detect possible sources of fire and smoke or even people in danger. This enables emergency crews to quickly get an overview of the extent and location of the fire by the transmitted video stream.

The operational area of the control center is a large rural area of approximately 7200 square kilometers in southern Brandenburg, which includes the city of Cottbus and the districts Dahme-Spreewald (LDS), Elbe-Elster (EE), Oberspreewald-Lausitz (OSL) and Spree-Neiße (SPN). This is visualized in Figure 1. To cover such a large area, decentralized storage of MAVs might be the most appropriate so that potential sources of fire can be reached quickly. To realize this, the MAVs are stored and maintained in purpose-built hangars, such as the one described by Heller et al. [3]. Such a hangar equipped with an affiliated UAV is illustrated in Figure 2. Thus, we consider an entire so-called Unmanned Aircraft System (UAS), consisting of several

UAV hangars, also known as drone-in-a-box systems, and an associated fleet of UAVs. The focus of this paper is on determining the appropriate resources (UAVs and hangars) and position them optimally in terms of covering the spatial demand to ultimately form a UAS configuration. For this, a Mixed-Integer Linear Programming (MILP) model is developed to solve the cooperative *Facility Location-Allocation Problem.*

The rest of the paper is structured as follows: Section 2 presents relevant literature, particularly in relation to UAS and location optimization for wildfire suppression. Section 3 presents the mathematical formulation of several strongly related MILP model variants for location-allocation optimization. This is followed by some computational experiments in the supply area of *Leitstelle Lausitz* in Section 4. Finally, Section 5 presents conclusions and ideas for future research directions.

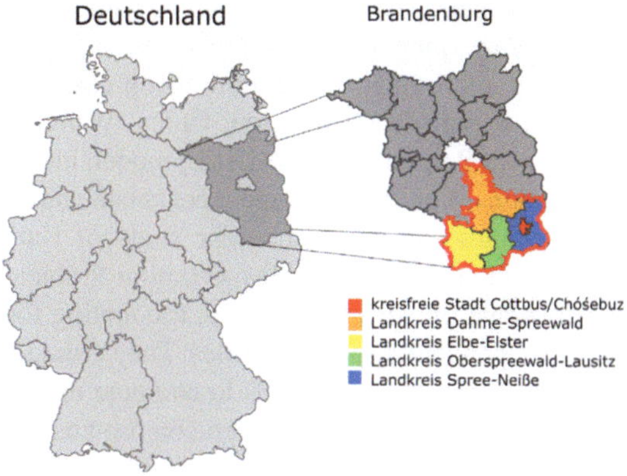

Fig. 1 The supply area of 7219 km^2 of control center *Leitstelle Lausitz* in Brandenburg, Germany [4]

Fig. 2 The *RescueFly* UAS consisting of UAV and hangar, developed for water rescue . (Photo Ralf Schuster, BTU)

2 Literature Review

The research field of wildfire fighting is very active due to its current relevance and urgency. Resource deployment, as well as the allocation of available resources, plays a critical role in wildfire fighting and is therefore being studied extensively. In particular, this involves deciding which resources should be stationed where and deployed when and to which location. To realize this for fire stations in urban areas, Murray [5] introduce a Complementary Threshold Coverage (CTC) model, which extends the classical Maximal Covering Location Problem (MCLP) [6] and Location Set Covering Problem (LSCP) [7]. The cost aspect of building new fire stations and maintaining existing stations is also taken into account, using the city of Elk Grove, California as an example. Similarly, Bao et al. [8] derive multiple MILP models for location-allocation of forest fire watchtowers from MCLP and LSCP formulations. They consider different requirements, namely minimizing cost while guaranteeing full coverage, maximizing coverage with a fixed budget and the bi-objective case of maximizing coverage while minimizing cost. The models developed were applied to a forest park in Guangzhou, China. Wang et al. [9] propose a Mixed-Integer Non-Linear Programming (MINLP) model, integrating budget restrictions, different size stations, a partial coverage mechanism and a cooperation mechanism for multi-type vehicles and perform a case study of Harbin, China. Rodriguez et al. [10] propose an emergency Facility Location Problem (FLP) MILP model maximizing the coverage of expected demand, accounting for demand types, vehicle types and region-dependent dispatching rules. They utilize an iterative procedure and a Hypercube Queueing Model (HQM) to compute the utilization of vehicles. Further, they introduce a case study on Concepción province, Chile.

As far as we know, there is no literature specifically on spatial location-allocation optimization of UAVs for wildfire fighting. In their studies, Belval et al. [11] found that, in general, decision support systems for the allocation of resources to fight wildfires have so far been insufficiently investigated. Still, there are a few publications on aerial firefighting resource location-allocation: Zeferino [12] proposes a MILP model to find good locations for aerial resources, distinguishing between initial attacks (IA) and extended attacks (EA) to suppress the fire. The model is inspired by a classical location model, the Maximum Expected Covering Location Problem (MEXCLP) [13]. It considers two types of coverage, provided by IA and EA, respectively, taking into account the availability of the aircraft. The objective of the model is to maximize the fire hazard coverage of both types. Helicopters and airtankers of different sizes as well as different types of airbases are considered. Spatial coverage of the supply area is incorporated by a circle indicating the operating range of the air vehicle. A case study of Portugal is included. Gustafsson et al. [14] developed a Binary Integer Programming (BIP) model to determine locations of aerial firefighting resources to increase the effectiveness of initial attacks for wildfire suppression. In contrast to Zeferino [12], they consider tactical, short-term, day-to-day dynamic resource location to be implemented and base their decisions on historical wildfire data. The model was applied to a use case in Sweden and compared to manual planning by experts. It turned out that the model delivered almost as good results as the experts, who, however, had access to far more information. However, their model is a fairly simple standard p-median model [15], which has some limitations, such as disregarding flight parameters, capacities and the occupancy of a potential site for no more than one vehicle. Chan et al. [16] consider observation UAVs and firefighting units (humans, trucks or UAVs) and develop a resource allocation strategy to maximize the coverage of a number of forest zones with assigned priority values and stochastically triggered fires of different intensity, measured by the fireline intensity. The allocation of the observation UAVs is thereby achieved by solving a Set Covering Problem (SCP). For a detailed general insight into location-allocation optimization approaches, we refer interested readers to the book of Laporte et al. [17].

In the following, we fill the identified gap by presenting a location-allocation MILP model that is specifically designed for UAS. Coverage is ensured in terms of time and remaining battery life. In addition, multi-vehicle and multi-stations are considered, which can have different sizes and capacities. Spatially varying demand is considered. Furthermore, cooperation between multiple UAVs is incorporated, and UAVs can also assume different tasks (e.g. monitoring, firefighting, searching).

3 Location-Allocation Optimization Model

We present MILP models that select from a variety of hangars, spatially locate them in a predefined supply area and equips them with a number of UAVs. The objective is to maximize the coverage of locally variable demand, depending on where there is a high risk of forest fires, by the allocated fleet of UAVs. For this, we adapted the approach of Zell et al. [18].

The basic concept of the model is that we allocate UAVs to UAV hangars and locate them at potential facility sites. UAVs can have different tasks or functions and need to be located to cover a spatially varying demand. This demand appears at discrete demand points. A demand point is considered covered if it can be reached within a certain time and additionally if the UAV fleet still has sufficient battery life to carry out various tasks at the demand point. For the battery-wise coverage, a cooperative covering location approach is chosen, introduced by Berman et al. [19], so that several UAVs contribute additively to reach the predetermined quantity of remaining battery life.

For the model formulation, we first define some sets. These include the sets of potential hangar locations, demand points, hangar types, UAV types and contribution types. A contribution type determines, how a vehicle contributes to the operation. This could be for example monitoring, extinguishing, medical equipment transportation, etc. The corresponding sets are listed in Table 1.

We model the capacity of facility sites, the capacity and delay of a hangar as well as the size, speed, acceleration rate, climb rate and battery capacity of a UAV. Furthermore, we introduce covering thresholds, a contribution matrix, a flight time tensor and signal strength tensors. All parameters are listed in Table 2. Note that

Table 1 Location-allocation optimization model sets

Symbol	Index	Description
$\mathcal{F}$	i	Set of potential hangar locations
$\mathcal{D}$	j	Set of demand locations
$\mathcal{U}$	u	Set of UAV types
$\mathcal{H}$	h	Set of hangar types
$\mathcal{C}$	c	Set of contribution or assignment types

the capacity of the hangar and the size of the UAV can be set arbitrarily. If the combinations in which different UAVs fit into the hangars are now to be determined, the parameters s_h and ψ_u must be set accordingly. To model location-allocation decisions we define some binary, integer and real decision variables. In particular, we need to decide where to locate a hangar of a certain type and how to equip the hangar with UAVs of a certain type. Further, we need to incorporate a mechanism to measure coverage of demand points with respect to approach time and battery capacity. Table 3 provides an overview of the decision variables.

We introduce the following model constraints:

$$\sum_{i \in \mathcal{F}, h \in \mathcal{H}} \mathbf{a}_{ih} \leq p, \tag{1}$$

$$\sum_{i \in \mathcal{F}} \mathbf{a}_{ih} \leq P_h^{max} \quad \forall h \in \mathcal{H}, \tag{2}$$

$$\sum_{i \in \mathcal{F}, h \in \mathcal{H}} \mathbf{n}_{uih} \leq N_u^{\max} \quad \forall u \in \mathcal{U}, \tag{3}$$

$$\sum_{u \in \mathcal{U}} \psi_u \mathbf{n}_{uih} \leq s_h \mathbf{a}_{ih} \quad \forall i \in \mathcal{F}, h \in \mathcal{H}, \tag{4}$$

$$\sum_{h \in \mathcal{H}} \mathbf{a}_{ih} \leq 1 \quad \forall i \in \mathcal{F}, \tag{5}$$

$$\sum_{h \in \mathcal{H}} s_h \mathbf{a}_{ih} \leq F_i^{max} \quad \forall i \in \mathcal{F}, \tag{6}$$

$$\sum_{u \in \mathcal{U}, i \in \mathcal{F}, h \in \mathcal{H}} \mathbf{n}_{uih}\, l_{uc}\, \phi_{uijh} \geq T_{jc} \mathbf{y}_{jc} \quad \forall c \in \mathcal{C},\ j \in \mathcal{D}, \tag{7}$$

$$\sum_{u \in \mathcal{U}, i \in \mathcal{F}, h \in \mathcal{H}} \mathbf{n}_{uih}\, l_{uc}\, \phi_{uijh} \geq T_{jc}^{\min} \quad \forall c \in \mathcal{C},\ j \in \mathcal{D}, \tag{8a}$$

$$\sum_{u \in \mathcal{U}, i \in \mathcal{F}, h \in \mathcal{H}} \mathbf{n}_{uih}\, l_{uc}\, \phi_{uijh} \geq T_{jc}^{\min} \mathbf{y}_{jc}^{\min} \quad \forall c \in \mathcal{C},\ j \in \mathcal{D}, \tag{8b}$$

Table 2 Location-allocation optimization model parameters. The parameters above the separation line are used directly in the MILP formulation, the parameters below the separation line are only used in pre-processing

Parameter	Domain	Description
p	$\mathbb{N}$	Number of hangars to be located
$F_i^{\max}$	$\mathbb{R}_{>0}$	Maximum capacity of a potential hangar site $i \in \mathcal{F}$
$P_h^{\max}$	$\mathbb{N}$	Number of available hangars of type $h \in \mathcal{H}$
s_h	$\mathbb{R}_{>0}$	Capacity of hangar $h \in \mathcal{H}$
ψ_u	$\mathbb{R}_{>0}$	Size of UAV $u \in \mathcal{U}$
$N_u^{\max}$	$\mathbb{N}$	Number of available UAVs $u \in \mathcal{U}$
l_{uc}	$[0, 1]$	Contribution of a UAV $u \in \mathcal{U}$ to assignment $c \in \mathcal{C}$
d_{jc}	$\mathbb{R}_{>0}$	Demand of type $c \in \mathcal{C}$ at demand point $j \in \mathcal{D}$
T_{jc}	$\mathbb{R}_{>0}$	Battery threshold to be reached at demand point $j \in \mathcal{D}$ per assignment $c \in \mathcal{C}$
$T_{jc}^{\min}$	$\mathbb{R}_{>0}$	Minimum battery threshold to be reached at demand point $j \in \mathcal{D}$ per assignment $c \in \mathcal{C}$
$\hat{T}_{jc}$	$\mathbb{R}_{>0}$	Time threshold to be reached at demand point $j \in \mathcal{D}$ per assignment $c \in \mathcal{C}$
$\hat{T}_{jc}^{\min}$	$\mathbb{R}_{>0}$	Minimum time threshold to be reached at demand point $j \in \mathcal{D}$ per contribution type $c \in \mathcal{C}$
ϕ_{uijh}	$\mathbb{R}_{>0}$	Signal strength values for UAV $u \in \mathcal{U}$ allocated to hangar $h \in \mathcal{H}$ located at $i \in \mathcal{F}$ emitting to $j \in \mathcal{D}$ to incorporate battery-wise coverage of demand points
C_{cuijh}	0, 1	Coverage tensor indicating if $j \in \mathcal{D}$ is time-wise covered from UAV $u \in \mathcal{U}$ allocated at hangar $h \in \mathcal{H}$ located at $i \in \mathcal{F}$
$C_{cuijh}^{\min}$	0, 1	Minimum coverage tensor indicating if $j \in \mathcal{D}$ is minimal time-wise covered from UAV $u \in \mathcal{U}$ allocated at hangar $h \in \mathcal{H}$ located at $i \in \mathcal{F}$
λ_1, λ_2	$\mathbb{R}_{<0}$	Coverage weighting coefficients
K	$[0, 1]$	Flight time weighting coefficient
τ_h	$\mathbb{R}_{>0}$	Delay of hangar $h \in \mathcal{H}$
$v_u^{\max}$	$\mathbb{R}_{>0}$	Maximum speed of UAV $u \in \mathcal{U}$
$a_u^{\max}$	$\mathbb{R}_{>0}$	Acceleration rate of UAV $u \in \mathcal{U}$
c_u	$\mathbb{R}_{>0}$	Climb rate of UAV $u \in \mathcal{U}$
σ_u	$\mathbb{R}_{>0}$	Battery capacity of UAV $u \in \mathcal{U}$
δ_{ij}	$\mathbb{R}_{\geq 0}$	Euclidean distance from $i \in \mathcal{F}$ to $j \in \mathcal{D}$
t_{uij}	$\mathbb{R}_{>0}$	Flight approach time of UAV $u \in \mathcal{U}$ from $i \in \mathcal{F}$ to $j \in \mathcal{D}$

$$\sum_{u \in \mathcal{U}, i \in \mathcal{F}, h \in \mathcal{H}} \mathbf{n}_{uih} C_{cuijh} \geq \hat{\mathbf{y}}_{jc} \quad \forall c \in \mathcal{C},\ j \in \mathcal{D}, \tag{9}$$

$$\sum_{u \in \mathcal{U}, i \in \mathcal{F}, h \in \mathcal{H}} \mathbf{n}_{uih} C_{cuijh}^{\min} \geq 1 \quad \forall c \in \mathcal{C},\ j \in \mathcal{D}, \tag{10a}$$

Table 3 Location-allocation optimization model variables

Symbol	Domain	Description
$\mathbf{n}_{uih}$	$\mathbb{N} \cup \{0\}$	number of UAVs of type $u \in \mathcal{U}$ stored in hangar of type $h \in \mathcal{H}$ located at facility $i \in \mathcal{F}$
$\mathbf{a}_{ih}$	$\{0, 1\}$	1 if hangar of type $h \in \mathcal{H}$ is sited at potential facility site $i \in \mathcal{F}$, 0 else
$\mathbf{y}_{jc}$	$\{0, 1\}$	1 if demand point $j \in \mathcal{D}$ is covered for contribution type $c \in \mathcal{C}$ w.r.t. battery threshold T_{jc}, 0 else
$\mathbf{y}_{jc}^{\min}$	$\{0, 1\}$	1 if demand point $j \in \mathcal{D}$ is covered for contribution type $c \in \mathcal{C}$ w.r.t. minimum battery threshold $T_{jc}^{\min}$, 0 else
$\hat{\mathbf{y}}_{jc}$	$\{0, 1\}$	1 if demand point $j \in \mathcal{D}$ is covered for contribution type $c \in \mathcal{C}$ w.r.t. time threshold $\hat{T}_{jc}$, 0 else
$\hat{\mathbf{y}}_{jc}^{\min}$	$\{0, 1\}$	1 if demand point $j \in \mathcal{D}$ is covered for contribution type $c \in \mathcal{C}$ w.r.t. minimum time threshold $\hat{T}_{jc}^{\min}$, 0 else
$\tilde{\mathbf{z}}_{cuijh}$	$\mathbb{R}_{>0}$	Auxiliary variables for $c \in \mathcal{C}, u \in \mathcal{U}, i \in \mathcal{F}, j \in \mathcal{D}, h \in \mathcal{H}$
$\tilde{\mathbf{z}}_{cuijh}^{\min}$	$\mathbb{R}_{>0}$	Auxiliary variables for $c \in \mathcal{C}, u \in \mathcal{U}, i \in \mathcal{F}, j \in \mathcal{D}, h \in \mathcal{H}$

$$\sum_{u \in \mathcal{U}, i \in \mathcal{F}, h \in \mathcal{H}} \mathbf{n}_{uih} C_{cuijh}^{\min} \geq \hat{\mathbf{y}}_{jc}^{\min} \quad \forall c \in \mathcal{C},\ j \in \mathcal{D}, \tag{10b}$$

$$\tilde{\mathbf{z}}_{cuijh} \leq \mathbf{n}_{uih} \quad \forall c \in \mathcal{C}, u \in \mathcal{U}, i \in \mathcal{F}, j \in \mathcal{D}, h \in \mathcal{H}, \tag{11}$$

$$\tilde{\mathbf{z}}_{cuijh} \leq M^{\text{loc}}\, \mathbf{y}_{jc} \quad \forall c \in \mathcal{C}, u \in \mathcal{U}, i \in \mathcal{F}, j \in \mathcal{D}, h \in \mathcal{H}, \tag{12}$$

$$\tilde{\mathbf{z}}_{cuijh} \leq M^{\text{loc}}\, \hat{\mathbf{y}}_{jc} \quad \forall c \in \mathcal{C}, u \in \mathcal{U}, i \in \mathcal{F}, j \in \mathcal{D}, h \in \mathcal{H}, \tag{13}$$

$$\tilde{\mathbf{z}}_{cuijh}^{\min} \leq \mathbf{n}_{uih} \quad \forall c \in \mathcal{C}, u \in \mathcal{U}, i \in \mathcal{F}, j \in \mathcal{D}, h \in \mathcal{H}, \tag{14}$$

$$\tilde{\mathbf{z}}_{cuijh}^{\min} \leq M^{\text{loc}}\, \mathbf{y}_{jc}^{\min} \quad \forall c \in \mathcal{C}, u \in \mathcal{U}, i \in \mathcal{F}, j \in \mathcal{D}, h \in \mathcal{H}, \tag{15}$$

$$\tilde{\mathbf{z}}_{cuijh}^{\min} \leq M^{\text{loc}}\, \hat{\mathbf{y}}_{jc}^{\min} \quad \forall c \in \mathcal{C}, u \in \mathcal{U}, i \in \mathcal{F}, j \in \mathcal{D}, h \in \mathcal{H}. \tag{16}$$

Further, we introduce the following objective functions:

$$\max \sum_{c,u,i,j,h} (d_{jc} - K(t_{uij} + \tau_h))\tilde{\mathbf{z}}_{cuijh}. \tag{17a}$$

$$\max \ \lambda_1 \sum_{c,u,i,j,h} d_{jc}\tilde{\mathbf{z}}_{cuijh} - K \sum_{c,u,i,j,h} (t_{uij} + \tau_h))\tilde{\mathbf{z}}_{cuijh} + \lambda_2 \sum_{c,u,i,j,h} d_{jc}\tilde{\mathbf{z}}^{\min}_{cuijh}. \tag{17b}$$

Now, we consider three different model formulations. The *Capacitated Cooperative Maximum Covering Location Problem (CCMCLP)* aims to maximize the weighted sum of battery and time covered demand points without considering minimal coverage. The second model variant that includes strict minimum coverage is named *Capacitated Cooperative Maximum Guaranteed Minimum Covering Location Problem (CCMGMCLP)*. The third model variant with weak minimum coverage is named *Capacitated Cooperative Maximum Weak Minimum Covering Location Problem (CCMWMCLP)*. The latter is intended to solve the problem that there is no feasible solutions for the CCMGMCLP if even a single demand point cannot be minimally covered. The models with associated equations are listed in Table 4. Equation (1) ensures that at most p hangars are placed. Equation (2) limits the number of hangars set up to the maximum number available. Analogously, Equation (3) limits the number of allocated UAVs to the maximum number available. Equation (4) ensures that the size of the allocated UAVs cannot exceed the hangar's capacity. Equation (5) ensures that at most one hangar is placed per site and Equation (6) ensures that only hangars of suitable size are placed at a site. Equation (7) incorporates a cooperative covering mechanism to cover the demand battery-wise: Only decision variables $\mathbf{y}_{jc}$ equal to 1 can contribute to the objective function, for which the total battery capacity of all UAVs at the demand point must exceed the corresponding threshold value. Equations (8a) and (8b) incorporate a cooperative minimum battery-wise covering mechanism, while the first is very strict and requires every demand point to be covered and the latter is more relaxed. Equation (9) incorporates an individual covering mechanism, time-wise. According to this, a demand point is only considered covered time-wise if it can be reached in a certain time, less than $\hat{T}_{jc}$, which is expressed by the tensor C_{cuijh}. Decision variable $\hat{\mathbf{y}}_{jc}$ can therefore only become 1 if coverage can be guaranteed from at least one selected site for the placed hangar and allocated UAV. Analogously, Equation (10a) and (10b) incorporate strict and weak minimum time-wise coverage. Equations (11) – (16) are auxiliary constraints for linearization. Using a big-M linearization approach, $\tilde{\mathbf{z}}_{cuijh}$ takes the value of $\mathbf{n}_{uih}\mathbf{y}_{jc}\hat{\mathbf{y}}_{jc}$ and $\tilde{\mathbf{z}}^{\min}_{cuijh}$ takes the value of $\mathbf{n}_{uih}\mathbf{y}^{\min}_{jc}\hat{\mathbf{y}}^{\min}_{jc}$. The resulting values are used in the objective functions from Equations (17a) and (17b). The first objective (17a) aims to maximize the weighted sum of time- and battery-wise covered demand points while minimizing the flight time, normalized through parameter K. The latter objective (17b) additionally maximizes the weighted sum of minimum time- and battery-wise covered demand points.

To compute flight times t_{uij}, signal strength values ϕ_{uijh} and coverage tensors C_{cuijh}, $C^{\min}_{cuijh}$, we utilize the UAS-specific parameters from below the separation line of Table 2 and consider euclidean distance δ_{ij} from $i \in \mathcal{F}$ to $j \in \mathcal{D}$. We divide the flight time t_{uij} of a UAV u from facility site i to demand point j into three intervals:

Table 4 Location-allocation optimization model variants with corresponding equations (constraints and objective)

Abbreviation	Name	Constraints	Objective
CCMCLP	Capacitated Cooperative Maximum Covering Location Problem	(1) – (7), (9), (11) – (13)	(17a)
CCMGMCLP	Capacitated Cooperative Maximum Guaranteed Minimum Covering Location Problem	(1) – (8a), (9), (10a), (11) – (16)	(17a)
CCMWMCLP	Capacitated Cooperative Maximum Weak Minimum Covering Location Problem	(1) – (7), (8b), (9), (10b) – (16)	(17b)

(1) The take-off from i to operation altitude ν with climb rate c_u, (2) the uniformly accelerated phase with acceleration rate $a_u^{\max}$ to maximum speed $v_u^{\max}$ and (3) the approach phase to j with maximum speed covering the rest of distance δ_{ij}. Utilizing the *kinematic equations of motion,* which describe the uniform acceleration from starting speed v_0 to target speed v_T for time t, distance d and acceleration a by

$$v_T = v_0 + at, \quad v_T^2 = v_0^2 + 2ad. \tag{18}$$

and $v_0 = 0$, $v_T = v_u^{\max}$, $a = a_u^{\max}$, $d = \delta_{ij}$, we obtain

$$t_{uij} = \frac{\nu}{c_u} + \frac{v_u^{\max}}{a_u^{\max}} + \frac{\delta_{ij} - \frac{(v_u^{\max})^2}{2a_u^{\max}}}{v_u^{\max}} = \frac{\nu}{c_u} + \frac{v_u^{\max}}{a_u^{\max}} + \frac{\delta_{ij}}{v_u^{\max}} - \frac{v_u^{\max}}{2a_u^{\max}} \tag{19}$$

$$= \frac{\nu}{c_u} + \frac{v_u^{\max}}{2a_u^{\max}} + \frac{\delta_{ij}}{v_u^{\max}}. \tag{20}$$

If distance δ_{ij} is too short for acceleration to maximum speed, phase (3) is omitted. Thus, we obtain

$$t_{uij} = \begin{cases} \frac{\nu}{c_u} + \frac{v_u^{\max}}{2a_u^{\max}} + \frac{\delta_{ij}}{v_u^{\max}}, & \text{if } \frac{(v_u^{\max})^2}{2a_u^{\max}} > \delta_{ij}, \\ \frac{\nu}{c_u} + \sqrt{\frac{2\delta_{ij}}{a_u^{\max}}}, & \text{else.} \end{cases} \tag{21}$$

The term *signal strength* was adopted from the concept of Berman et al. [19], as it represents a monotonically decreasing function in distance, just like a signal that becomes progressively weaker as the distance traveled grows. In our case, the signal strengths can be interpreted as remaining battery time of a UAV u after approaching the target demand point j from a hangar h located at potential facility site i. This approach is designed to ensure time coverage while incorporating a cooperative mechanism as explained in detail by Berman et al. [19]. Note that with $2\,t_{uij}$ we consider both outbound and return time and this approach could be modified by defining emergency landing sites to have more remaining battery left by avoiding the long return flight. Thus, signal strength ϕ_{uijh} is computed by

$$\phi_{uijh} = \phi(t_{uij}, \tau_h, \sigma_u) := \max\{\sigma_u - 2t_{uij} - \tau_h, 0\} \quad \forall u \in \mathcal{U}, h \in \mathcal{H}, i \in \mathcal{F}, j \in \mathcal{D}. \tag{22}$$

The coverage tensors in Equations (9), (10a) and (10b), intended to ensure time coverage, contain binary values only, depending on surpassing time thresholds $\hat{T}_{jc}$ and $\hat{T}_{jc}^{\min}$, respectively:

$$C_{cuijh} = \begin{cases} 1, & \text{if } l_{uc}(\tau_h + t_{uij}) \leq \hat{T}_{jc} \\ 0, & \text{else,} \end{cases} \quad \forall u \in \mathcal{U}, h \in \mathcal{H}, i \in \mathcal{F}, j \in \mathcal{D}, \tag{23}$$

$$C_{cuijh}^{\min} = \begin{cases} 1, & \text{if } l_{uc}(\tau_h + t_{uij}) \leq \hat{T}_{jc}^{\min} \\ 0, & \text{else,} \end{cases} \quad \forall u \in \mathcal{U}, h \in \mathcal{H}, i \in \mathcal{F}, j \in \mathcal{D}. \tag{24}$$

4 Computational Experiments

We now apply the location-allocation model presented in the previous chapter to a use case in Brandenburg, Germany. To do this, we use the whole supply area of the regional control center *Leitstelle Lausitz* as the operational area and start with defining local demand points and potential hangar facility sites. To measure the demand, we need to identify where potential wildfires occur. We choose a fast and reproducible approach by using open data in the form of wildfire-prone areas in OpenStreetMap (OSM) [20]. We also fully include the areas that are only partially located in the supply area. This is reasonable, as the control center may be asked for help in neighboring areas. The selected features and associated OSM keys and values as well as related wildfire risk assessments are listed in Table 5 and visualized in Figure 3. We only consider high-risk features (cp. "Selection" column in Table 5) to maintain computational efficiency but this approach can be refined if necessary, also through other data sources, such as regional historical deployment data. By adjusting the features used to determine the demand, it is possible to customize the scope of the UAS application. For example, one could consider extending the scope of the UAS from wildfire fighting to fire fighting in general by including residential areas. To calculate the specific coordinates and weighting values of the discrete demand points, we use the centroid of the area and calculate the size of the area, respectively. However, this results in an excessively large number of 18, 597 demand points. This would make solving the MILP far too time-consuming, so we utilize k-means clustering to reduce the number of demand points. The k-means clustering is applied by proximity to the centroid using Euclidean distance, and each cluster is represented by its centroid. There are certainly enough potential hangar sites in the entire supply area. However, in order to get a first initial overview of where the hangars should be stationed and to keep the computational effort under control, we use an equidistant grid approach to determine the locations. The spatial demand distribution is visualized as a heat map in Figure 4. Table 6 shows the parameter configuration used for the instances, which are solved below using CCMCLP, CCMGMCLP and CCMWMCLP. The missing parameters from Table 2, namely ϕ_{uijh}, C_{cuijh}, $C_{cuijh}^{\min}$ are calculated from Equations (22), (23) and (24).

The solution of the *CCMCLP* and the *CCMWMCLP* for $p = 9$ hangars to be located, $|\mathcal{D}| = 1000$ clustered demand points and $|\mathcal{F}| = 23$ potential facility sites, resulting from a spacing of 15, 000 meters, are shown exemplary in Figures 5 and 6. Some minor differences in the exact position of the hangar are visible. These

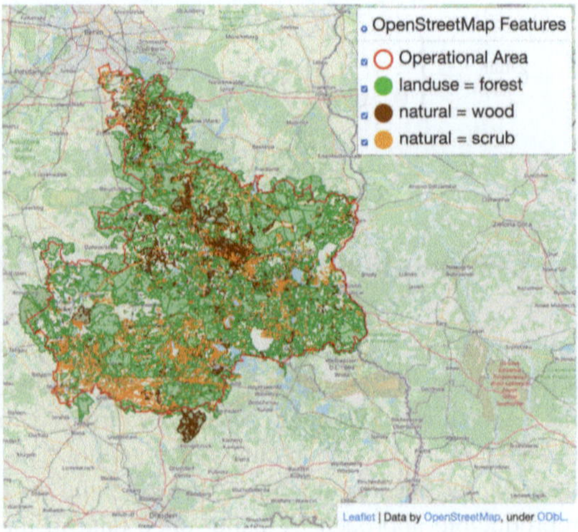

Fig. 3 OpenStreetMap (OSM) features for fire-prone areas in the operational area (red outline) in Southern Brandenburg, Germany

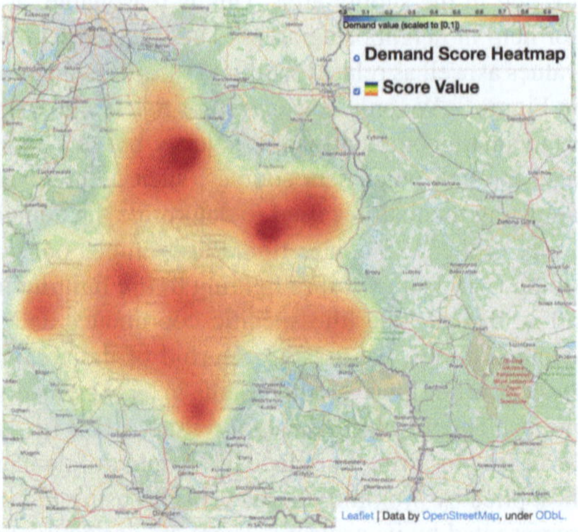

Fig. 4 Demand heatmap for fire-prone areas in Southern Brandenburg, Germany

Table 5 Selected OpenStreetMap (OSM) features with respective keys, values and wildfire risk assesments to identify wildfire-prone areas. The information was obtained from the OSM Wiki [21] and in consultation with local experts. The "Selection" column shows the considered features to solve the location-allocation problem

Key	Value	Description	Wildfire risk	Selection
landuse	forest	Land managed for forestry	Very high	Yes
	farmland	Farmland used for tillage	High	No
	meadow	Meadow or pasture	High	No
	grass	Mown and managed grass	Moderate	No
natural	wood	Tree areas	Very high	Yes
	scrub	Uncultivated land covered with shrubs, bushes or stunted trees	Very high	Yes
	heath	Dwarf-shrub habitat	High	No
	grassland	Grass dominated areas	High	No
	wetland	Natural areas subject to inundation or with waterlogged ground	Moderate	No

Table 6 Parameter configuration for solving various location-allocation optimization models

Name	Parameter	Value		
		CCMCLP	CCMGMCLP	CCMWMCLP
Maximum capacity of a potential hangar site	$F_i^{\max}$	$3\,\forall i \in \mathcal{F}$	$3\,\forall i \in \mathcal{F}$	$3\,\forall i \in \mathcal{F}$
UAV size	ψ_u	2	2	2
Available UAVs	$N_u^{\max}$	8	8	8
Hangar capacity	s_h	3	3	3
Available hangars	$P_h^{\max}$	8	8	8
Hangar delay	τ_h	$8s$	$8s$	$8s$
Operation altitude	ν	$50.0m$	$50.0m$	$50.0m$
UAV battery capacity	σ_u	$1800s$	$1800s$	$1800s$
Battery threshold	T_{jc}	$1200s\,\forall j \in \mathcal{D}$	$1500s\,\forall j \in \mathcal{D}$	$1500s\,\forall j \in \mathcal{D}$
Minimum battery threshold	$T_{jc}^{\min}$	–	$1080s\,\forall j \in \mathcal{D}$	$1080s\,\forall j \in \mathcal{D}$
Time threshold	$\hat{T}_{jc}$	$100s\,\forall j \in \mathcal{D}$	$60s\,\forall j \in \mathcal{D}$	$60s\,\forall j \in \mathcal{D}$
Minimum time threshold	$\hat{T}_{jc}^{\min}$	–	$120s\,\forall j \in \mathcal{D}$	$120s\,\forall j \in \mathcal{D}$
UAV maximum speed	$v_u^{\max}$	$25.0\frac{m}{s}$	$25.0\frac{m}{s}$	$25.0\frac{m}{s}$
UAV acceleration rate	$a_u^{\max}$	$5.0\frac{m}{s^2}$	$5.0\frac{m}{s^2}$	$5.0\frac{m}{s^2}$
UAV climb rate	c_u	$5.0\frac{m}{s^2}$	$5.0\frac{m}{s^2}$	$5.0\frac{m}{s^2}$

adjustments are necessary to account for minimal coverage for as many demand points as possible as the CCMCLP simply ignores points that are not covered, but the CCMWMCLP at least considers minimal coverage of the demand points that cannot be fully covered. Note that for the CCMGMCLP there is no solution for $p \leq 23$, since no minimum coverage can be guaranteed for all points. This means that a feasible solution cannot even be produced using all 23 potential hangar locations. It is therefore extremely difficult to completely minimally cover the extremely large coverage area, where covering here means being able to reach every demand point in at most $\hat{T}_{jc}^{\min} = 120$ seconds while guaranteeing a remaining battery life of $T_{jc}^{\min} = 1080$ seconds. For other values of $\hat{T}_{jc}^{\min}$ and $T_{jc}^{\min}$, this would not apply and the analysis would have to be performed again.

In the second use case, we have two different types of UAVs, one for localization and monitoring and the other for extinguishing fires. Further, we consider two

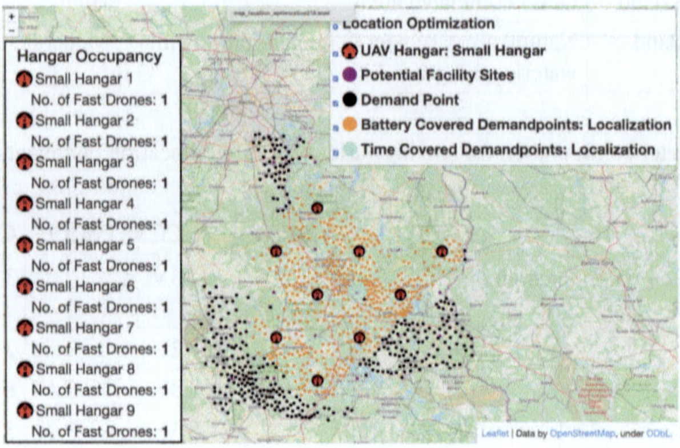

Fig. 5 Optimal solution of the CCMCLP for the first use case and $p = 9$ hangars to be located

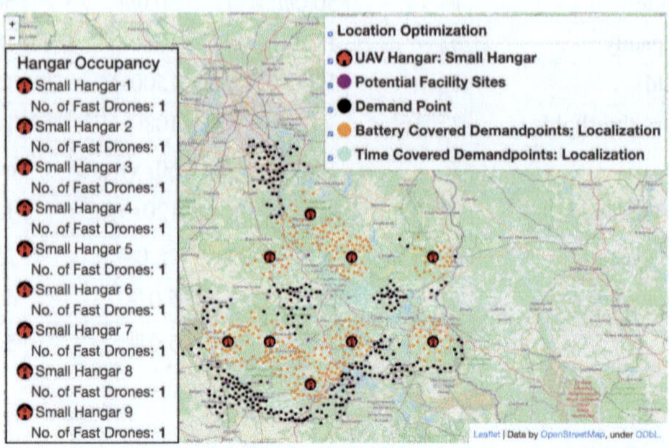

Fig. 6 Optimal solution of the CCMWMCLP for the first use case and $p = 9$ hangars to be located

different hangar types, a very large one, which provides capacity for up to two extinguishing UAVs or up to four monitoring UAVs, and a small one, which offers space for one extinguishing UAVs or two monitoring UAVs. The associated parameters do not correspond to a specific type of UAV or hangar, as we do not have precise data such as battery life, speed, etc. for extinguishing UAVs, for example. Nevertheless, the parameters have been chosen realistically. They are listed in Table 7. The instances were solved for $p = 9$ using state-of-the-art commercial solver IBM ILOG CPLEX [22] version 21.1.1.0 on an Apple MacPro with an Intel Xeon W running 32 threads parallel at 3.20 GHz clock speed and 768 GB RAM an run into the time limit of 48 hours. A MIP gap of 13.90% and 26.00% was achieved for the CCMCLP and CCMWMCLP, respectively. This indicates that scaling effects can be quite severe for larger instances. The results are visualized in Figures 7 and 8. The figure clearly shows that the short battery life is a major limitation for the extinguishing drones.

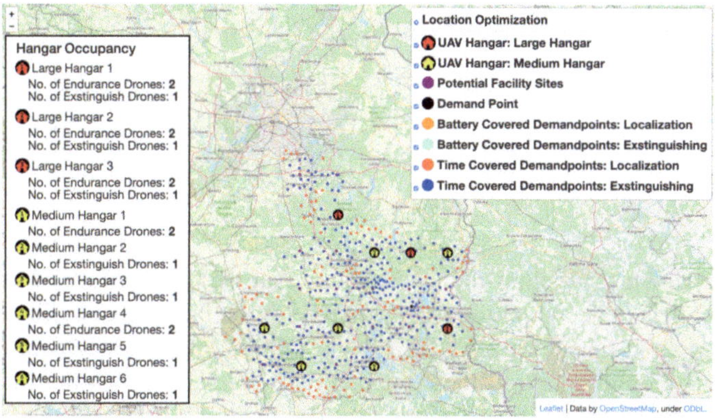

Fig. 7 Optimal solution of the CCMCLP for the second use case and $p = 9$ hangars to be located

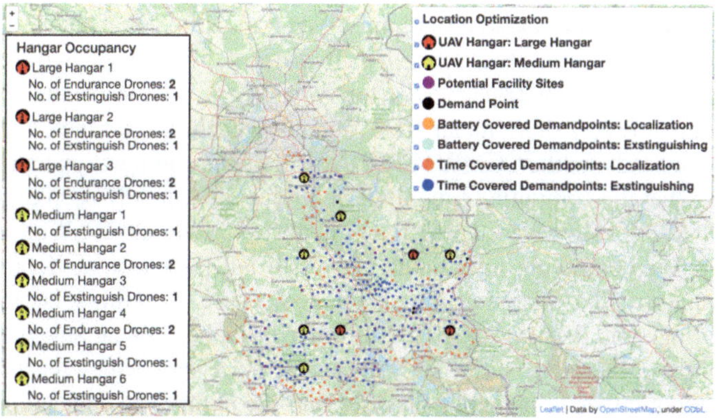

Fig. 8 Optimal solution of the CCMWMCLP for the first second case and $p = 9$ hangars to be located

Table 7 Parameter configuration of the second use case for CCMCLP and CCMWMCLP. Note that the '[]'- notation of the values indicates multiple types (e.g. of UAVs or hangars)

Name	Parameter	Values
Maximum capacity of a potential hangar site	$F_i^{\max}$	$4 \, \forall i \in \mathcal{F}$
UAV size	ψ_u	[1, 2]
Available UAVs	$N_u^{\max}$	[10, 10]
Hangar capacity	s_h	[2, 4]
Available hangars	$P_h^{\max}$	[7, 3]
Hangar delay	τ_h	$[10.0s, 15.0s]$
Operation altitude	ν	$120.0m$
UAV battery capacity	σ_u	$[3300s, 1500s]$
Battery threshold	T_{jc}	$[600s, 120s] \, \forall j \in \mathcal{D}$
Minimum battery threshold	$T_{jc}^{\min}$	$[120s, 30s] \, \forall j \in \mathcal{D}$
Time threshold	$\hat{T}_{jc}$	$[600s, 720s] \, \forall j \in \mathcal{D}$
Minimum time threshold	$\hat{T}_{jc}^{\min}$	$[1200s, 1200s] \, \forall j \in \mathcal{D}$
UAV maximum speed	$v_u^{\max}$	$[82.80\frac{m}{s}, 50.0\frac{m}{s}]$
UAV acceleration rate	$a_u^{\max}$	$[5.0\frac{m}{s^2}, 3.0\frac{m}{s^2}]$
UAV climb rate	c_u	$[6.0\frac{m}{s^2}, 3.0\frac{m}{s^2}]$

In the following, we show the investigation that we have performed on how the value of p affects the computing time required for the first use case. To do this, we have solved an instance of the CCMCLP and CCMWMCLP for each of the values $p = 1, \ldots, 23$ and analyzed the computation times, as well as the MIP gaps if an instance could not be solved within the specified time limit. All instances were solved with CPLEX version 21.1.1.0 [22] on an Apple MacPro with an Intel Xeon W running 32 threads parallel at 3.20 GHz clock speed and 768 GB RAM with relative MIP gap set to 1% and a time limit of 10800 seconds (or 3 hours). The results of the examination are summarized in Figures 9 and 10. Note that the MIP gap is only displayed if the time limit was exceeded. If the time limit is adhered to, the MIP gap is guaranteed to be below 1%. For low values of $p \leq 4$, the computation time remains comparatively short, under 1000 seconds in each case for the CCMCLP and under 3000 seconds for the CCMWMCLP. After that, the computation time increases abruptly, so that the computing time limit of 3 hours has already been exceeded for the CCMWMCLP. However, the MIP gap is always below 20 %. With increasing values of p, the computing time finally seems to fall back to a moderate level. If more than half of all available potential facility sites are to be filled, i.e. for $p \geq 12$, the computing time decreases rapidly. In order to make reliable statements about the scalability of the models, further analyses are necessary, which must include in particular the number of potential sites $|\mathcal{F}|$ and the number of demand points $|\mathcal{D}|$. Furthermore, the computing times dependency on the number of hangar types $|\mathcal{H}|$, UAV types $|\mathcal{U}|$ and contribution types $|\mathcal{C}|$ must also be investigated.

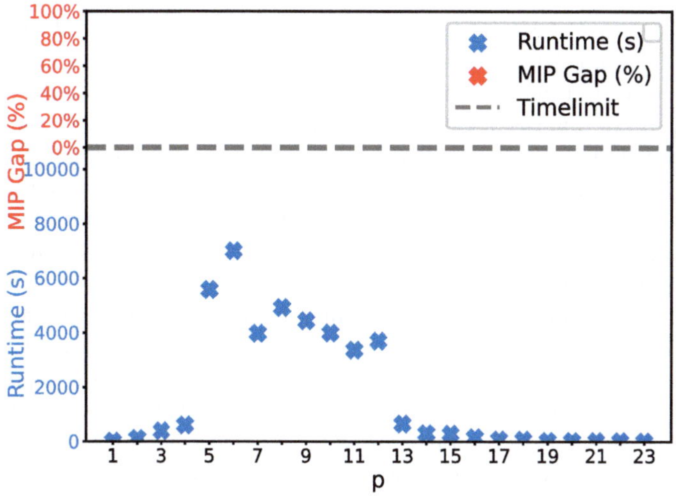

Fig. 9 Runtime and MIP gap of CCMCLP solution for $p = 1, \ldots, |\mathcal{F}|$ with $|\mathcal{F}| = 23$

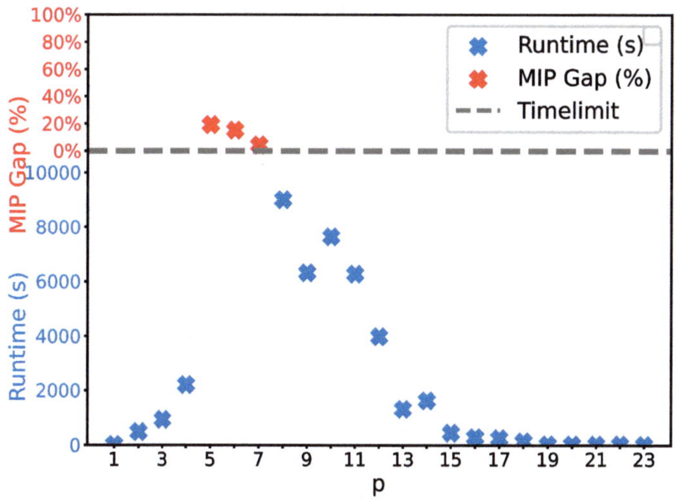

Fig. 10 Runtime and MIP gap of CCMWMCLP solution for $p = 1, \ldots, |\mathcal{F}|$ with $|\mathcal{F}| = 23$

5 Conclusions

In this paper, we introduced location-allocation MILP models for a UAS consisting of UAVs allocated to UAV hangars for the purpose of first response for wildfire fighting. We presented the mathematical formulation for three slightly different models, namely the *Capacitated Cooperative Maximum Covering Location Problem (CCMCLP)*, the *Capacitated Cooperative Maximum Guaranteed Minimum Covering*

Location Problem (CCMGMCLP) and the *Capacitated Cooperative Maximum Weak Minimum Covering Location Problem (CCMWMCLP)*. The models include multi-type hangars with different capacities and multi-type UAVs of different sizes and maximum speeds. Further, a UAV can contribute to various objectives and coverage is incorporated time-wise and battery-wise. To cover a demand point battery-wise, multiple UAVs can contribute cooperatively. This means that the remaining battery life of all UAVs is added together. The mathematical formulation of the different variations of the proposed MILPs and their differences and purposes were explained. Finally, we applied the models to a use case in Brandenburg, Germany. The intention was to equip an entire supply area of a control center (*Leitstelle Lausitz*) with the proposed UAS while guaranteeing a certain level of coverage. Our model revealed that many more hangars than expected are needed to cover the entire supply area of over 7000 square kilometers at least for a minimum threshold. However, the models form a good basis for decision-making of the best possible use of these resources (UAVs and hangars), which might be limited due to budget constraints. Altogether, our numerical experiments indicate that the models presented are a valuable tool for location-allocation optimization for fighting wildfires despite possible scaling problems for larger instances that induce large computation times.

We envisage a number of extensions. One further research idea would be to carry out an extensive sensitivity and calculation time analysis for various model parameters. Since the output of the models also depends heavily on some model parameters, such as the selected threshold values or weighting parameters, an iterative approach could be considered to determine optimal parameters. A further idea is applying a simulation approach, similar to Zell et al. [18], to simulate UAS-assisted wildfire fighting operations. This would allow us to estimate the average arrival time of the UAVs so we can compare different UAS configurations and quantify the benefits of acquiring extra resources. These models assume that only one incident occurs at a time and that all resources are available at all times. As a result, some use cases may exceed the limits of the model, e.g. deliberately set fires with multiple ignition points or particularly forest fire-prone regions with multiple simultaneous fires. This requires certain extensions of the model, e.g. similar to MEXCLP, a certain employment rate could be incorporated for the vehicles in order to account for large workloads at peak times, such as in very dry summer months or as a result of the increasing number of wildfires caused by climate change. Alternatively, a backup coverage mechanism could be considered. Further investigations can also include the development of a fully automated framework, similar to the approach from Zell et al. [18], performing location-allocation optimization, optimal UAV flight trajectory optimization for the target areas, simulating rescue operations and utilizing automated image detection of smoke, people or sources of fire via neural networks.

Acknowledgements The authors express their gratitude to Leitstelle Lausitz for their cooperation as well as the whole *RescueFly* and *AIRCIS* project groups for fruitful collaboration and inspiration.

Competing Interests The authors have no conflicts of interest to declare that are relevant to the content of this chapter.

Funding This study was funded by the *Federal Ministry for Digital and Transport Germany* through the research projects *RescueFly* (grant number 45ILM1016C) and *AIRCIS* (grant number 19FS2039C).

Data availability The software code used in this work is freely available in the GitHub repository *uas-firefly* at https://github.com/SaschaZell/firefly-uas or on PyPI at https://pypi.org/project/firefly-uas/.

References

1. Schmidt, J., Fügenschuh, A.: Planning inspection flights with an inhomogeneous fleet of micro aerial vehicles. In Cottbus Mathematical Preprints 21 (2021). https://doi.org/10.26127/BTUOpen-5656
2. Wulf, M., Jahn, U., Meier, K., Radtke, M.: Tree species composition of a landscape in North-Eastern Germany in 1780, 1890 and 2010. Forestry Int. J. Forest Res. **90**(2), 174–186 (2016). https://doi.org/10.1093/forestry/cpw061
3. Heller, A., Harradi, R., Hardt, W.: Hangar system for unmanned aerial vehicle autonomous missions. In 2024 International Symposium ELMAR, pp. 291–294, Zadar, Croatia (2024). https://doi.org/10.1109/ELMAR62909.2024.10694237
4. Leitstelle Lausitz. Über uns. , n.d. Online. https://www.leitstelle-lausitz.de/leitstelle/ueber-uns/ accessed 30 Oct 2024
5. Murray, A.T.: Optimising the spatial location of urban fire stations. Fire Safety J. **62**, 64–71 (2013). Special Issue on Spatial Analytical Approaches in Urban Fire Management. https://doi.org/10.1016/j.firesaf.2013.03.002
6. Church, R., ReVelle, C.: The maximal covering location problem. Papers of the Reg. Sci. Assoc. **32**(1), 101–118 (1974). https://doi.org/10.1007/BF01942293
7. ReVelle, C., Toregas, C., Falkson, L.: Applications of location set-covering problem. Geogr. Anal. **8**, 65–76 (1976). https://doi.org/10.1111/j.1538-4632.1976.tb00529.x
8. Bao, S., Xiao, N., Lai, Z., Zhang, H., Kim, C.: Optimizing watchtower locations for forest fire monitoring using location models. Fire Safety J. **71**, 100–109 (2015). https://doi.org/10.1016/j.firesaf.2014.11.016
9. Wang, J., Liu, H., An, S., Cui, N.: A new partial coverage locating model for cooperative fire services. Inf. Sci. **373**, 527–538 (2016). https://doi.org/10.1016/j.ins.2016.06.030
10. Rodriguez, S.A., De la Fuente, R.A., Aguayo, M.M.: A facility location and equipment emplacement technique model with expected coverage for the location of fire stations in the concepci6n Province, Chile. Comput. Ind. Eng. **147**, 106522 (2020). https://doi.org/10.1016/j.cie.2020.106522
11. Belval, E.J., Short, K.C., Stonesifer, C.S., Calkin, D.E.: A historical perspective to inform strategic planning for 2020 end-of-year wildland fire response efforts. Fire **5**(2), 35 (2022). https://doi.org/10.3390/fire5020035
12. Zeferino, J.: Optimizing the location of aerial resources to combat wildfires: a case study of Portugal. Nat. Hazards **100**(3), 1195–1213 (2020). https://doi.org/10.1007/s11069-020-03856-6
13. Daskin, M.S.: A maximum expected covering location model: formulation, properties and heuristic solution. Transp. Sci. **17**(1), 48–70 (1983). https://doi.org/10.1287/trsc.17.1.48
14. Gustafsson, V.S., Granberg, T.A., Waldemarsson, M.: Using optimization modeling to determine locations for aerial firefighting resources based on wildfire risk data. In Proceedings of the 21st ISCRAM Conference, vol. 21, Münster, Germany (2024). https://doi.org/10.59297/bvpahd97
15. Daskin, M.S., Maass, K.L.: The p-Median problem. In Location Science, pp. 21–45. Springer, Cham, Switzerland (2015). https://doi.org/10.1007/978-3-319-13111-5_2

16. Chan, H., Tran-Thanh, L., Viswanathan, V.: Fighting wildfires under uncertainty: A sequential resource allocation approach. In Proceedings of the Twenty-Ninth International Conference on International Joint Conferences on Artificial Intelligence, pp. 4322–4329, Yokohama, Japan (2021). https://doi.org/10.24963/ijcai.2020/596
17. Laporte, G., Nickel, S., Saldanha-da Gama, F.: Introduction to Location Science. Springer Nature, Cham, Switzerland (2019). https://doi.org/10.1007/978-3-030-32177-2_1
18. Zell, S., Schneidereit, T., Fügenschuh, A., Breuß, M.: Advanced search and rescue operations for drowning swimmers using autonomous unmanned aircraft systems: Location optimization, flight trajectory planning and image-based localization. In Cottbus Mathematical Preprints 31 (2024). https://doi.org/10.26127/BTUOpen-6866
19. Berman, O., Drezner, Z., Krass, D.: Discrete cooperative covering problems. J. Oper. Res. Soc. **62**(11), 2002–2012 (2011). ISSN 1476-9360. https://doi.org/10.1057/jors.2010.176
20. OpenStreetMap contributors. Openstreetmap (2024). https://www.openstreetmap.org. Accessed 9 November 2024
21. OpenStreetMap contributors. OpenStreetMap Wiki: Main Page , n.d. https://wiki.openstreetmap.org. Accessed 11 November 2024
22. International Business Machines Corporation. IBM ILOG CPLEX Optimizer Reference Manual (2023). https://www.ibm.com/docs/en/icos/22.1.1

Correction to: Stochastic Optimization for Scheduling Thinning Operations in Mediterranean Pine Forest Stands Under Fire Damage Risk

M Acuna, MA Varo-Martínez, V Lerma-Arce, G Palacios-Rodríguez, and R Navarro-Cerrillo

Correction to: Chapter 1 in: F. Alvelos et al. (eds.), *Advances in Optimization and Wildfire*, Lecture Notes in Operations Research, https://doi.org/10.1007/978-3-032-03108-2_1

The original version of this chapter was published with incorrect Reference and Citations. Chapter Reference corrections are carried out in Reference section [4, 7, 8, 24] respectively.

The updated version of this chapter can be found at
https://doi.org/10.1007/978-3-032-03108-2_1

F. Alvelos et al. (eds.), *Advances in Optimization and Wildfire*, Lecture Notes in Operations Research, https://doi.org/10.1007/978-3-032-03108-2_16

MIX
Papier aus verantwortungsvollen Quellen
Paper from responsible sources
FSC® C105338

If you have any concerns about our products,
you can contact us on
ProductSafety@springernature.com

In case Publisher is established outside the EU,
the EU authorized representative is:
Springer Nature Customer Service Center GmbH
Europaplatz 3, 69115 Heidelberg, Germany

Printed by Libri Plureos GmbH
in Hamburg, Germany